THE PUBLIC EVALUATION OF GOVERNMENT SPENDING

edited by
G. Bruce Doern
and
Allan M. Maslove

Proceedings of a Conference
sponsored by the
Institute for Research on Public Policy
and the
School of Public Administration
Carleton University
October 19-21, 1978
Ottawa

Institute for Research on Public Policy/Institut de recherches politiques

Montreal
1979

ISBN 0 920380 19 0

Legal Deposit First Quarter
Bibliothèque nationale du Québec

2nd Printing, February 1980

Institute for Research on Public Policy/Institut de recherches politiques
2149 MacKay St.
Montreal, Quebec H3G 2J2

Preface

In October 1978 the School of Public Administration of Carleton University and the Institute for Research on Public Policy co-sponsored a "Conference on Methods and Forums for the Public Evaluation of Government Spending" in Ottawa. This meeting was one of a series of "National Conferences on Governmental Processes" being co-sponsored by the IRPP.

This volume includes the major papers and three selected commentaries delivered at the Conference. The papers are presented here essentially as they were presented to the Conference; only minor editorial changes have been made.

In addition, the first essay in this volume presents a discussion of the major themes that emerged during the course of the Conference. It is not intended as a summary of the discussion although it does note the central arguments raised by various participants whose contributions are not included here. Rather, it attempts to assess the prospects for public expenditure evaluation and identify the key areas for future reform.

As with our other conferences, we hope the reader finds that some light has been shed on an important and much discussed topic.

Michael J.L. Kirby
President
March 1979

Préface

En octobre 1978, L'Ecole d'Administration publique de l'université Carleton et l'Institut de recherches politiques tenaient conjointement une conférence sur les méthodes et les tribunes appropriées à l'évaluation collective des dépenses gouvernementales, à Ottawa. Cette rencontre s'inscrivait dans une série de conférences nationales sur les processus gouvernementaux organisées, en partie, sous l'égide de l'IRP.

Ce volume comprend les principaux documents présentés lors de la conférence ainsi que trois commentaires choisis. Les textes sont reproduits ici, pour l'essentiel, tels qu'ils ont été soumis à la conférence, à l'exception de quelques changements mineurs aux fins d'édition.

De plus, le premier essai de cet ouvrage présente une analyse des principaux thèmes qui ont été soulevés au cours de la conférence. L'article ne se veut pas un résumé des discussions, bien qu'il relève les principaux arguments formulés par divers participants dont les interventions sont absentes de notre ouvrage, mais plutôt une appréciation des perspectives d'évaluation des dépenses publiques, identifiant les éléments clés d'une réforme éventuelle.

Nous espérons que le lecteur puisera dans ce livre, comme dans toutes nos conférences, quelque information valable sur un sujet important et très controversé.

Michael J.L. Kirby
Président
Mars 1979

Acknowledgement

We would like to acknowledge and thank Trish Donnelly of the School of Public Administration and Donald Wilson of the IRPP for their organization and administration of the Conference.

Acknowledgement

Table of Contents

Chapter One

The Public Evaluation of Government Spending: From Methods to Incentives

by
G. Bruce Doern

and
Allan M. Maslove

It is said that "money talks." It can also be said that public money speaks with a forked tongue, or—to extend our metaphorical binge—that public money sings with a thousand voices. At first glance, one might employ these metaphorical summaries of the October 1978 Conference on Methods and Forums for the Public Evaluation of Government Spending. Sponsored by the School of Public Administration at Carleton University and the Institute for Research on Public Policy, the conference brought together about 150 persons from across Canada drawn from all levels of government, and from business, unions, interest groups and academe. This book presents a collection of the major conference papers and commentaries. The purpose of the editors' introductory chapter is to highlight some of the issues raised at the conference and to critically examine recent trends in public expenditure evaluation which the conference itself in part reflected.

The conference was convened at a time when there was apparently a new urgency to a very old subject—namely, how and through what public processes and forums can one evaluate and understand both the aggregate growth and levels of government spending and the effectiveness and efficiency of particular governmental programs and activities. The latter half of the 1970s have witnessed an unprecedented degree of criticism and cynicism about government in general and the conference reflected a good part of this concern. At the same time, the conference also reflected a genuine view that change was both necessary and possible.

The views about change, however, were characterized by a kind of intellectual and practical sobriety that had not characterized the earlier wave of methodological and evaluative reform in the 1960s, the wave that brought PPB, management by objectives, systems analysis, and the like. The conference showed that any evaluation reformer of the 1980s will be greeted as part missionary and part used-car salesman. This healthy ambivalence was reinforced by the issues raised by the other dimension of the conference title, namely, the various notions of what "public" evaluation meant and the several forums in which it might occur. For some, "public" simply meant more openness in government. For others, "public" meant the broader Canadian community and the citizen-voter in his or her various forms. Commentators such as Professor Pierre-Paul Proulx stressed regional decentralized "publics" and forums. The various forums for public evaluation similarly included such diverse arenas as Parliament, the media, interest groups, academe, and other governments within the Canadian federal system.

Each of the papers and commentaries in this volume reflect this cautious ambivalence about expenditure reform and evaluation. In this introductory commentary, however, we will highlight four themes which emerged at the conference and which seem to us to be especially important for an understanding of the long-run dimensions of and prospects for public expenditure evaluation. We will first examine the marked shift in focus from the 1960s' preoccupation with methodology to the current concern with why such methods have fallen into disrepute or have yielded little change. In the second section of the chapter we will focus on the "self-interest" motivated actors and the incentives and disincentives they face in the conduct and use of evaluation in each of several of the main forums or institutions, such as the executive, Parliament and the media. The third section will attempt to demonstrate how the prospect of future reform rests on understanding these incentives and on designing new public, especially non-governmental forums. Finally, we will examine the degree to which the more recent wave of reforms must be judged against the need to strike an acceptable democratic balance between the capacity of governments to govern (that is, to mobilize political power) and the need to constrain the excesses of power and to foster a healthy sense of precariousness among democratically elected leaders. Such a balance must be struck, moreover, in the context of developments which suggest that there is a need for a longer time frame for the making of many public decisions.

METHODOLOGY: FROM OPTIMISM TO REALISM

The papers in this volume each reflect in their own way a sense of modesty about the problems of and the prospects for expenditure evaluation. There is little enthusiasm for the quick "technical fix" which seemed so often to characterize the early emergence of "rational" policy analysis in the 1960s. Professor **Irwin Gillespie**'s analysis clearly stresses the limitations of technical methods. His discussion of the Gladstonian, Keynesian, and 'Ksanian voter (the latter an ingeniously Canadian name for the "socialist" voter whose prime concern is redistribution) shows starkly the mixture of norms that will colour any evaluation and its interpretation. Relative preferences for efficiency, income stability, and redistribution will be present along with other norms more particular to specific kinds of expenditure. Thus, Gillespie concludes that the analyst's role in the whole evaluation of public spending debate is a technical and limited one.

The paper by **Harry Rogers**, Canada's recently appointed Comptroller General, also stresses the need for practicality and realism. He characterizes the role of his office as a catalyst "for more efficient, effective and responsive administration of government expenditures." He stresses that his office will not attempt to implement "grand and expensive systems thinking that they will provide easy cures to diverse and complex problems." He cites his own sobering experience in the private sector where, in the latter 1960s the systems' advocates also bit off more than they could chew. The Comptroller General's comments reveal a search for a more precise definition of "program" and of whether all programs are evaluable. In several respects the strategy is one of the "soft sell" of program evaluation in an era of expenditure restraint.

Evolution of Different Budgetary Systems

Aaron Wildavsky's comparative analysis of the recent evolution of different budgetary systems and of the deeply embedded survival instincts of the traditional (and usually disparaged) line-item budget serves as further notice about the need for sobriety. He stresses the different purposes, temporal assumptions, and, hence, conflicts between different kinds of budgetary control systems. Above all, he lodges his analysis in the behavioural milieu in which budgets are perceived, especially in public bureaucracies. Future program evaluators will, for example, have to confront the following logical assertion of Wildavsky about evaluating by program:

> Imagine one of us deciding whether to buy a tie or a kerchief. A simple task, one might think. Suppose, however, that organizational rules require us to keep our entire wardrobe as a unit. If everything must be rearranged when one item is altered, the probability we will do anything is low. The more tightly linked the elements, and the more highly differentiated they are, the greater the probability of error (because the tolerances are so small), and the less the likelihood error will be corrected (because with change, every element has to be recalibrated with every other one that was previously adjusted.) Being caught between revolution (change in everything) and resignation (change in nothing) has little to recommend it.
>
> Program budgeting increases rather than decreases the cost of correcting error. The great complaint about bureaucracies is their rigidity. As things stand, the object of organizational affection is the bureau as serviced by the usual line-item categories from which people, money, and facilities flow. Viewed from the standpoint of bureau interests, programs to some extent are negotiable; some can be increased and others decreased while keeping the agency on an even keel or, if necessary, adjusting it to less happy times, without calling into question its very existence. Line-item budgeting, precisely because its categories (personnel, maintenance, supplies) do not relate directly to programs, are easier to change. Budgeting by programs, precisely because money flows to objectives, makes it difficult to abandon objectives without abandoning the organization that gets its money from them. It is better that

> non-programatic rubrics be used as formal budget categories, thus permitting a diversity of analytical perspectives, than that a temporary analytic insight be made the permanent perspective through which money is funnelled.
>
> The good organization is interested in discovering and correcting its own mistakes. The higher the cost of error—not only in terms of money but also in personnel, programs and prerogatives—the less the chance anything will be done about them. Organizations should be designed, therefore, to make errors visible and correctible, that is noticeable and reversible, which in turn is to say, cheap and affordable.[1]

Wildavsky argues that it is always necessary to think about objectives but it is not always logical to design budgets and evaluations around them, at least not if one is interested in seeing change actually occur. The Wildavsky view is often greeted with utter bewilderment by those with very abstract views about rational decision making, and who fail to understand the nature of political behaviour and the rationality of such behaviour. Political and program objectives not only involve different rankings of values by different groups and beneficiaries at any one point in time, but they also shift over time, often over very brief periods of time. Much of political conflict is a conflict among goods.

Programs and Objectives

The analysis of public policy shows the persistence of policy inconsistency. Departments are often subject concurrently to several statutory commands and political orders. Many programs are supported to pursue different objectives including some conflicting objectives. Often much of what passes for the so-called "administration" or "implementation" of a program is merely another arena in which earlier political differences are re-examined. Individuals and groups who "lost" at the early stage of policy development rejoin the battle. Often the coinage of political exchange changes. Differences are now expressed in terms such as "we lack adequate staff," or "the regulations are unfair," or "we will need much more time," whereas at earlier stages the differences may have been expressed in terms of actual objectives or values. Living with policy inconsistency is a fact of political life which expenditure evaluators will have to confront. Moreover, groups and institutions with a stake in programs will always want to know in whose interests evaluation is being conducted.

EVALUATION INCENTIVES AND DISINCENTIVES

The conference papers and discussion demonstrated the increasing degree to which it is recognized that self-interest motivation and incentives fundamentally affect the behaviour of both institutions and actors in the

---

[1] See pp. 68-69 of this volume.

public accountability process. The structure of incentives in several forums is simply not conducive to the conduct and utilization of the kind of expenditure evaluation envisaged by many reformers. Indeed, there exists a dominant system of disincentives for evaluation.

Role of the Media

Dian Cohen's commentary on the media's (especially the press') role in public expenditure evaluation, while anecdotal, shows the degree to which the short attention span of the press, the pressure to get the story out, the rotation of career assignments, and often the lack of knowledge act as disincentives for ongoing evaluation. She asserts, moreover, that when, on occasion, a journalist does become especially knowledgeable he or she, as often as not, is likely to be co-opted by a Cabinet minister and joins that minister's political staff. At best there exists only a small handful of Canadian journalists who have the knowledge and the degree of interest to devote time to expenditure evaluation, particularly on individual programs.

Lack of Incentives

Michael English's analysis of expenditure scrutiny in the United Kingdom and the conference commentaries by **James Gillies**, M.P. and **Maurice Leclair**, Secretary to the Treasury Board, confirm other recent analyses by Dobell[2] and Thomas[3] which show the lack of positive incentives for the conduct and use of evaluation by parliamentary institutions. This is especially the case in the Canadian House of Commons where the strength of partisan political norms, party discipline, and adherence to the view of Parliament as gladiatorial struggle create at best only a selective interest in evaluation. Much of the incentive structure can be traced to the central place where rigid rules of want of confidence and party discipline are imposed on the parliamentary accountability process.

Executive Power

The predominance of executive power, reinforced by the laws and traditions of Cabinet and administrative secrecy on the one hand, and information policies and calculated leaks of information on the other, serves to insure the weakness of parliamentary committees and other modes of inquiry. All of this is aided and abetted by the historical dominance of one political party at the federal level and the ambivalance of opposition strategists and leaders about the proper role of the opposition. The latter

---

[2] Peter Dobell, 1977.
[3] Paul Thomas, 1977.

ambivalence oscillates between "opposition for the sake of opposition," to "constructive opposition," to opposition as "alternative future government." There is, moreover, a strong tendency, as James Gillies stressed, for opposition strategy during Question Period to be significantly influenced by the press and media coverage of the day. Hence the incentives of the media reinforce the incentives of the parliamentary arena.

It must be stressed that not all the incentives of Parliament operate against expenditure evaluation. The parliamentary struggle is often a forum for good old-fashioned political intelligence and evaluation in which alternative views about expenditure levels and program norms are asserted and which influence governmental choices, often just by the *anticipation* of parliamentary criticism. There are, moreover, examples of quite effective program evaluations conducted by the Senate Finance Committee on such programs and activities as manpower policy, Information Canada, and the accommodation program of the Department of Public Works.

Existence of Disincentives

That strong disincentives for meaningful evaluation also exist in the executive and bureaucracy is also increasingly obvious. Wildavsky's comparative analysis makes the bureaucratic disincentive abundantly clear. The bureaucratic incentive structure disproportionately rewards spending, the proposing of new programs, and relatively short-term pay-offs. It negatively influences longer term programs and planning and the production of information necessary for conducting evaluations. Moreover, when data do exist there clearly is an incentive to distort or to keep them from "outsiders" who may use them to an agency's disadvantage; in the bureaucracy, information is a major currency of power.

Michael Prince's analysis of evaluation units in the Canadian federal government adds further dimensions to the dilemma.

> It is not difficult to understand why there are few resources or organizations in government departments dedicated to the evaluation of established programs. When policy advisory groups evaluate government programs they are assessing the *raison d'être* of departments. Government departments and agencies derive their support, strength, and significance from programs. Hence, a thorough and investigative evaluation can be viewed as a threat to program personnel. A government department's attitude toward the utility of policy analysis is also affected by the senior officials and whether they are sympathetic towards analysis, the nature of the policy or program, the political strength of the program's clientele, and the extent to which the department has relied in the past on analytic and evaluative methods.
>
> Even where resources are allocated to groups for program evaluation, many policy advisors believe that the evaluation role conflicts with their other functions and their need, particularly during the early years of operation, to develop good relations with departmental line officials. The technically easier and organizationally less threatening functions like policy development have therefore been emphasized by advisors at the expense of program evaluation. In some groups the performance of a fire-fighting role, whether intended or not, has dominated and in some cases prevented the

development of planning and evaluation roles. This problem cannot be simply solved by keeping groups separate from the heat of day-to-day concerns.

Other constraints and considerations confronting groups in organizing their advisory roles include insufficient resources (time, staff, position classifications, and facilities), difficulties in recruiting and retaining qualified personnel, the inexperience of some advisors, the absence of relevant data, and technical problems in the use of analytic methods. Also, the distinctive characteristics of policy advisors have influenced staff-line relations and group effectiveness. There was undoubtedly a negative reaction from some departmental managers and officials to the introduction of policy-advisory groups because the groups were staffed by "whizz kids." Furthermore, the groups' role was often not clear to, or accepted by, line officials. Lacking support and access to channels of influence and persuasion, some groups have remained outside the mainstream of departmental policy and management processes.[4]

Academic Incentives

As for academic incentives, Irwin Gillespie's emphasis on the limitations of methodological technique and on the primacy of competing values and norms is probably not the majority view in academic circles. The incentive system of academe still is more prone to applaud evaluative technique and the search for causal knowledge. The growing role of contract research and consulting opportunities for those academics with skills and interests in policy and program evaluation can also lead to its own forms of co-optation and to the absence of both the substance and the appearance of independence. On another level, academic comment tends to be of limited value for public evaluation because of the rigidities of disciplines and because academic communication is more likely to occur in the rarefied climate of learned refereed journals, rather than in other kinds of media and publication.

As will hopefully be clear in the next section of this introductory essay, the intention in the above criticism of the incentive systems is not to argue naively that each of these institutions should become transformed in total. The dominant incentive systems in each arena or forum are sustained over time because they help fulfill other intended and valued functions of these institutions. A failure to understand the perverse consequences of these incentives on public evaluation can, however, result in naive reform proposals that can directly lead to the making of policy and public choices which are themselves influenced, in part at least, by strategies designed to avoid direct evaluation. Several of the papers in this book both stress and provide examples of this fact.

---

[4] Michael Prince, 1979.

Tax Expenditures

Professor **Allan Maslove**'s examination of tax expenditures show how a heretofore relatively hidden instrument of policy—the use of the tax system to confer benefits—might be increasingly preferred by decision makers over other direct, more visible instruments (such as regular expenditures) precisely because they are less subject to evaluation and the display of such choices in public documentation. Maslove's initial evaluation of Canadian tax expenditures shows their pronounced regressive pattern, and he recommends that the Finance minister should be required to publish data about such benefits, a practice now required in the United States.

Bureaucratic Growth

In a somewhat different vein, the analysis of bureaucratic growth by Professors **Richard Bird** and **David Foot** illustrates both the need for and difficulties in utilizing empirical evaluation. Much of the instinctive business and corporate criticism of public expenditure growth as reflected in the commentary at the conference by **William Twaits**, former President of Imperial Oil, is implicitly or explicitly imbedded in beliefs and/or assertions about the growth of bureaucracy and the expansionary appetites of bureaucrats. While the Bird and Foot paper is careful not to imply any single causal links between bureaucratic growth and expenditure growth, it does address a number of myths about the parameters of bureaucratic growth. The analysis shows the need to differentiate the various definitions of "public employment" as opposed to "civil service" employment, and points to the relatively greater rate of growth of public employment in the early post-war period.

Both the analyses of tax expenditures and of bureaucratic growth point to the need for and value of careful public evaluation. That such "analyses" will be screened through various ideological blinkers can be confidently predicted. Ideological assumptions and beliefs can be both a substitute for further thought, and/or a short-hand expression of a cluster of values. Reports about the claimed end of ideology in public debate are greatly exaggerated.[5] Ideology may not always determine policy but it almost always helps to foreclose the *range* of options considered and the way in which formal evaluations will be received and handled. Particular program areas, moreover, may be especially sensitive to the triggering of both grand ideologies (left-right) and of somewhat narrower but still pervasive paradigms or ideas (e.g., the value accorded the work ethic in several social programs).[6] Thus, for example, one response to tax expenditure evaluation

[5] Daniel Bell, 1970.

[6] R. Manser, 1975, and Peter Aucoin, 1979

may be that tax expenditures should not be defined as taxes the government chooses not to collect because that implies all resources are state property, but rather as taxes that the government has no business collecting. Similarly, subtle evaluations of bureaucratic growth will not convince anyone who simply believes government has grown "too much."

The inherently political basis of evaluation is stressed in Professor **Hugh Armstrong**'s commentary. Evaluation of the public sector is inherently selective and occurs within the pervasive existence of *inequality* both in the opportunities by various communities to evaluate and to gain access to political authorities, and in the actual distribution of benefits of governmental activity (regulatory and fiscal). Hence, it is not surprising, in Armstrong's view, that governments do not look first to evaluating the redistributive consequences of programs.

The general tenor of the conference papers makes it abundantly clear that, although some technical problems about evaluation exist, it is the issue of incentives and interests which goes to the heart of future reform strategies.

REFORM AND THE DESIGN OF NEW INCENTIVES

That there are no simple cures and no single "correct" results in the public expenditure evaluation process is compellingly clear. The key task is to try to devise and implant new incentives which on the whole will induce new or different patterns of behaviour in which open evaluation will be viewed by those who are the objects of evaluation to be somewhat less threatening to their own survival and, hence, become a motor for change where change is warranted. It is also important to build on the improvements that did occur in the wake of the heady 1960s' wave of analytical reform. This earlier reform period did, as **Dean Crowther**'s review of American experience shows, help generate "better numbers" and in many quarters generated a begrudging acknowledgement of the need for analysis. In the Canadian context, future reform can build on the earlier phases of the development of the "policy analysis" industry. The emergence of the industry can be traced to the mid-1960s when central policy-advisory bodies such as the Science Council and Economic Council were created and began publishing their reports. This was quickly followed in the late 1960s by the emergence of small "planning" units in the central agencies. Central agency officials in turn encouraged the development of departmental policy-advisory groups whose numbers grew rapidly in the early 1970s. Central agency officials also spearheaded the experimentation with policy ministries such as the Ministry of State for Science and Technology and the Ministry of State for Urban Affairs whose initial base of influence was to be knowledge and analytical capacity. The policy analysis industry also took on somewhat less of a government flavour because of the emergence and strengthening in the mid-1970s of several private or independent policy "think tanks" such as the

C.D. Howe Research Institute, the Institute for Research on Public Policy, the Conference Board in Canada, the Fraser Institute, the Canada West Foundation, and the Ontario Economic Council.

As Doern and Aucoin point out:

> The proposals to institutionalize program evaluation systems in all federal departments were in part a reaction against, and in part a continuation of, the development of the policy analysis industry. It was a reaction against it in that the program evaluation movement was aimed more pragmatically at ensuring that current programs were operating effectively and efficiently. This pragmatism was encouraged by the clarion calls of the Auditor General to achieve better "value for money", as well as by a growing view that earlier, broader efforts at policy analysis had produced groups that were almost like a university without students. Scepticism about the development of a new version of "paralysis by analysis" was not unfounded . . . [but] there emerged both benefits and costs. A more competitive policy-advisory process was created which promoted more open avenues of exchange and stimulation between central agencies and departments, between line and staff. Many new people with excellent minds were brought into government.[7]

The need to think concretely about incentives is perhaps most urgently necessary when considering the spate of reform proposals which have emerged in recent years. These include, among others:

a) the development and publication of a White Paper on public expenditure which would present the government's medium-term (three to five-year) expenditure projections and plans

b) the passage of sunset laws, whereby organizations and programs would automatically lapse unless positively re-enacted following an evaluation of its performance

c) the development of zero-based budgeting and systematic ongoing internal program evaluation and

d) the passage of public information laws and other reform measures to make access to bureaucratically generated and monopolized information more readily available to the public and other institutions of government.

White Paper on Expenditure Projections

The idea that the federal government should be required to publish and present to Parliament three to five-year expenditure projections has been suggested by several individuals and groups.[8] Modelled partly on existing British practice, a Canadian expenditure White Paper process could involve the following stages:

— publication by the President of the Treasury Board of an annual White Paper on Government Expenditure

---

[7] Doern and Aucoin, 1979

[8] G. Bruce Doern, 1978, and Hal Kroeker, 1978.

— creation of a central Parliamentary Committee on the Expenditure Budget and the Economy to receive and scrutinize the White Paper and to invite briefs and testimony on the White Paper and
— full parliamentary debate on the White Paper and the Expenditure Committee's evaluation of it.

The White Paper could be prepared and published by the Treasury Board and/or the Finance Department. It would be based partly on the existing internal program forecasts that are already produced (but not published) by the Treasury Board.

The White Paper would contain information on expenditure plans and projections for a three-year period, on a departmental and program basis, and on the basis of economic categories of expenditure. The document would also project revenues on a year-by-year basis for the three-year period and present information on the government's medium-term economic forecast and the assumptions behind it.

The President of the Treasury Board would present the White Paper to Parliament late in each calendar year before thc annual estimates are presented to Parliament. This could eventually result in a two-day parliamentary debate, following the receipt by the House of Commons of an initial evaluation and commentary by the proposed Committee on the Expenditure Budget and the Economy.

In addition to preparing an initial commentary on the White Paper, this committee could prepare other reports and receive other public testimony from interested groups and from expert bodies, such as the Economic Council of Canada, and from private bodies, such as the C.D. Howe Research Institute (among others). Such a committee should be adequately staffed to undertake these responsibilities.

In terms of parliamentary scrutiny and accountability, it is argued that such a process would create a forum which would induce a collective opposition response on total expenditure levels, and would also by definition extend the time frame of analysis. Parliamentary influence would be seen to operate only where it effectively can, namely on medium-term *future* expenditure priorities. That it has little or no influence on current year expenditures seems obvious. In terms of the *internal* process of accountability within the executive it is argued that the White Paper process would help break down interdepartmental boundaries because, in contrast to current practice where program forecasts and other expenditure data are held close to the vest by the central agencies, departments would know better and more openly where they stood precisely because the projections and data base would be published.[9]

---

[9] G. Bruce Doern, 1977.

To understand fully how such an expenditure reform might fare, however, one must relate it to the evolution and politics of government spending and budgeting alluded to by Aaron Wildavsky.

Political Attitudes and Government Spending

That government financial management and accountability are influenced greatly by the prevailing political attitudes toward government spending seems patently obvious. In Canada, and in other Western countries, there appear to have been three distinct phases of attitude and approach to government expenditure. From the mid-1930s to the late 1950s, despite a great growth in the number and range of government functions, concern for frugality and regularity of spending and financial management was predominant. The yearly rhythm of budgetary activity centred on the annual estimates (which were largely categorized under standard categories of expenditure, such as salaries or supplies).

Beginning in the early 1950s and particularly during the 1960s, the dominant influence was John Maynard Keynes. Attitudes to government spending were influenced by viewing government expenditure as an instrument of demand management in fiscal policy making. Thus, to the early concern for frugality and regularity were added the requirements of Keynesian economics.

Beginning in the 1960s with the Glassco report in Canada and the Plowden and Fulton reports in the United Kingdom there emerged a third set of attitudes to the expenditure process. This phase was dominated by concern for positive "program" management at a time when governmental expenditure programs were growing rapidly. The phase was also characterized by efforts to extend the time frame of expenditure planning within government beyond the one-year estimates period to a three to five-year period. Budget estimates, moreover, were categorized by a mixture of output-oriented program units tied to program objectives as well as by the traditional standard categories of expenditure.

The expenditure process (and hence ultimately financial management) was thus successively influenced by three kinds of values or desired end results. Each is understandable on its own, but taken together they have tended to confuse and divert the attentions of ministers, senior public servants, as well as major economic groups. Any future reforms of the expenditure process, including the proposed White Paper on Expenditure and its review, will have to keep in mind these frequently conflicting values about public expenditure.

Degrees of Control

Temporal as well as value conflicts also arise over different *degrees* of control over public expenditure of different kinds. Control may be more or less effective because of different political jurisdictions (e.g., provincial governments in Canada) or may vary according to the kind of government agency or entity (e.g., Crown corporations). Control may differ for capital as opposed to current spending, for demand-related non-discretionary (or locked-in statutory) spending as opposed to discretionary spending, expenditure with different product and labour market impacts (e.g., construction costs), and expenditure suitable for short-term demand management manipulation as opposed to that which is not.

The British experience, as well as the realities of Canadian government and politics, suggests strongly that implementing a process involving the publication and review of a White Paper on Government Expenditure will encounter rough passage, at least initially, and perhaps always. There will be resistance both in the executive and in the parliamentary area for all the same reasons suggested by the British experience. One can, in addition, see that a federal government might resist such a proposal entirely if provincial governments were not similarly prepared to go out on a public limb with their expenditure plans and projections. It is also highly likely that the nature of the public debate generated would be one carried out among experts, since only a handful of financial journalists and/or academics might have the time and interest to engage in such debate.

All reforms are supported or opposed by a variety of interests for quite divergent reasons. The White Paper could be seen by some interests as a convenient device to induce and perpetuate political and fiscal conservatism in that it would require the government to relate revenues to expenditure in a more integrated fashion. It might, however, also induce at least an "embarrassment factor" as government would be confronted yearly with the unexplained gaps between the expenditure projections of three years previous and their current spending levels. Other interests might see such a reform as a useful device to understand and to educate others about expenditure trends. Such public data might show how or where specific social spending programs are starved for funds and, hence, show more clearly and regularly the winners and losers in the "redistribution" of resources alluded to earlier.

Sunset Laws

Sunset laws have been adopted in several American states and have been suggested for use in Canada.[10] The sunset law has been seen largely as a device for the reform of regulatory bodies, but in fact it could apply to any government department, agency, or Crown corporation. A sunset proposal was adopted at the Ontario Conservative Party annual meeting in May 1976, and if enacted would require provincial regulatory boards to justify their existence every two years or be dissolved. Supporters of the proposal saw it

[10] Robert D. Behn, 1977.

as a way to lessen governmental intervention and to reduce the growth of public bureaucracies. A similar concept was suggested in a federal Progressive Conservative Party policy paper published in 1978.[11]

Sunset laws can be seen as one attempt to give greater visibility to regulatory phenomena. Regulation is a more hidden dimension of government, in that regulation, both in the aggregate and at the margin, has not been readily converted into the common denominator of expenditure, especially its effect on private budgets (individual and corporate). The movement to identify "tax expenditures" can be seen as an effort to make such choices more visible. Sunset laws, however, seem to be based on heroic assumptions about parliamentary capability to evaluate such bodies and programs in the short-run crowded parliamentary agenda. The experience of revising the *Bank Act* every ten years may be indicative in this context.

Zero-Based Budgeting and Program Evaluation

Proposals to adopt zero-based budgeting and program evaluation are a reflection of similar concerns and conflicts as those discussed above. Advocates of zero-based budgeting envisage a process whereby *all* spending (in Canadian terms, both the A and B budgets) would be evaluated every year with "new" money increasingly generated by the cancellation or alteration of activities judged to be inefficient or unnecessary.[12] Wildavsky's analysis ought to leave us under no illusions as to the predictable fate of this reform.

Freedom of Information Laws

Another area of reform seems essential, although the conference dealt with it only indirectly. This reform involves greatly strengthened freedom of information laws. Such laws are not in themselves panaceas since they can lead to their own versions of an information overload, but there can be little doubt that the Canadian system of Cabinet parliamentary government is excessively secretive. While most people now at least pay lip-service to the idea of freer public access to bureaucratic information, few appear to realize the potential changes in the incentive system that such laws would bring. Perhaps more than any other single reform discussed in this volume or elsewhere, public access to information has the most potential to alter bureaucratic incentives in favour of more effective public evaluation. The possibility (threat) of public scrutiny would fundamentally affect policy formulation, administration, and evaluation. The strong resistance that meets any real attempts at reform in this direction should be, in itself, an indicator

---

[11] *Globe and Mail*, May 17, 1976, and Progressive Conservative Party of Canada, 1978.

[12] Peter A. Pyhrr, 1977.

that secrecy is, in many ways, the lynchpin of the present incentive system and of the potential changes in incentives that public access could bring.

The issue thus comes down to not evaluation techniques and results because there are no uniquely "correct" results (as Gillespie makes clear), but to incentives and processes for evaluation. We have stressed earlier the fact that restructuring the incentives of major institutions to induce better public evaluation of government spending will not be easy. It would thus seem essential that reforms be selected with particular attention paid to the potential of any reform to induce changes in behaviour. The media, Parliament, interest groups, and other public forums all have a system of incentives which has evolved over long periods of time to fulfill numerous important functions, only some of which can be said to deal with the evaluation of aggregate or particular expenditure activity. In the light of this, it seems to us to be especially necessary that universities and other potentially independent centres of knowledge such as policy institutes have special responsibilities. One might be more confident about the assumption of such responsibilities by these "knowledge" centres if there was greater evidence of a willingness and capacity to communicate as opposed to merely publish information. Academics are still more prone to publish for and to address their colleagues than they are to communicate to a wider public. Recent competition from newly emerging private policy institutes may induce new communicative habits.

Public evaluations by academe and private policy institutes still encounter the classic "free rider" problem. Public evaluators are a public good and, hence, one can presumably get them without paying for them. Conference speakers such as **Carl Beigie** of the C.D. Howe Research Institute and **David Slater** of the Economic Council of Canada both addressed the question of who should conduct and publish evaluation of expenditure and stressed the difficulties of acquiring funding. General evaluations are by definition not in any individual group's particular interest and, hence, they are reluctant to support such "public" evaluative enterprises.

THE MODERN DEMOCRATIC BALANCE

The public evaluation of government spending requires a healthy balance between at least two kinds of knowledge, namely, good old-fashioned political intelligence—the intelligence of elected politicians—and the knowledge which emanates from clear publicly communicated research and analysis. It is clear that in the achievement of this knowledge balance several institutions have a role to play. These institutions, Parliament, media, interest groups, community interests, and academe, are part of an accountability chain, a chain which in Canada has many weak links. It is not sufficient, however, to strengthen the chain by adding on a "wish list" of reforms. In addition to their individual effect on incentives, these reforms must be assessed *collectively* in relation to other essential elements of the democratic balance, namely, the balance between ensuring that governments can govern and can mobilize power on the one hand, and that the excesses of power can be constrained on the other. Both the knowledge balance and the balance of power must be struck, moreover, at a time when conflicts are arising not only about the several ends to which policy and programs can be devoted, but also about the length of *time frames* for public choice.

Conflict Over Time Frames

Several factors have influenced the conflict over time frames. First, the rational aspirations of central agency policy reformers suggested that a longer term view of things was necessary. Second, the conflict between the inherent medium-term realities of economic "supply" management and the shorter term "fine tuning" requirements of economic Keynesian demand management is increasingly evident. The explicit need for "medium-term" federal-provincial economic summit meetings has been recognized. Third, analyses of the regulatory process show the unprecedented concern about the long-term aggregate consequences of growing government regulation. Finally, we have also seen demands for more Berger-style inquiries to insure that a thorough and long-term view of major socio-economic investments are properly assessed by those most affected. As Doern and Aucoin point out:

> On the one hand, there is a compelling logic for such a development and extension of the time frame for policy. Yet there is an understandable reluctance in a liberal democratic federal state to want to call, or admit the need, for a more "planned" economy. At the same time, one observes an increasing volume of demands in the short-run, leading in part to a view that government is overloaded and incapable of responding. Assuming finite and scarce resources, the political system is forced either to induce a lowering of expectations in the population to reduce demands; to respond substantively to selected demands and say no to the majority of others; or to devise a more elaborate array of responses including several combinations and permutations of the following: express symbolic concerns, establish consultative forums, study the problem (task force, royal commission, hearings, references to think tanks), visibly reorganize agencies/programs, and develop new expenditure and/or regulatory programs.
>
> It is the increasing virtuosity in the use of the latter array of responses that led to numerous reform proposals in the late 1970s.[13]

Each of the reforms discussed briely in the previous section is in part based on a concern for assessing longer term consequences. But each reform in the context of examining *longer term* consequences and extending the time frame of policy crowds the short-term decision agenda. Reforms can be seen to encourage a longer term view of things and the need for more analysis, but

---

[13] Doern and Aucoin, 1979.

each imposes new short-term lags and delays and, hence, induces greater uncertainty about the future. The situation becomes ripe for abuse. Analysis and research can too easily become a substitute for more substantive responses to policy problems, and can serve interests whose objective may be to prevent *any* government action. Analysis is sought for numerous reasons and to support and/or frustrate several causes. Policy analysts and evaluators have a vested interest in promoting more analysis. They are part of the knowledge business. Central agency-line department (not to mention federal-provincial) analytical competition can easily cross that invisible line beyond which it ceases to be organizationally or politically stimulating, and becomes counter-productive and even destructive of co-operative action. That there are limits to the value of the public expenditure evaluation game is increasingly obvious. But one should not throw the analytical baby out with the bathwater because it was not that long ago that no *public* policy evaluation capability, save that of periodic royal commissions, existed in Canada.

Chapitre un

L'évaluation collective des dépenses publiques: des méthodes aux stimulants

par
G. Bruce Doern et *Allan M. Maslove*

"L'argent est éloquent" dit-on. D'autre part, on peut ajouter que les deniers publics ont la langue fourchue ou—sans trêve de métaphore— qu'ils ont mille et une voix. De prime abord, ce coup d'oeil métaphorique à la conférence d'octobre 1978 sur les *Méthodes et tribunes appropriées à l'évaluation collective des dépenses publiques* se justifie. Cette conférence, sous les auspices de l'Ecole d'Administration publique de l'université Carleton et de l'Institut de recherches politiques, a réuni quelque 150 personnes provenant de toutes les régions du pays et représentant tous les paliers de gouvernement, le secteur des affaires, les syndicats, les groupes d'intérêt particuliers et le milieu universitaire. Cet ouvrage rassemble les principaux travaux de la conférence et les commentaires qui y ont été prononcés. Les éditeurs se proposent dans ce chapitre d'introduction de mettre en lumière quelques-uns des problèmes qu'abordait la conférence et de procéder à l'examen critique des tendances récentes de l'évaluation des dépenses publiques, comme la conférence elle-même l'a fait ressortir.

La conférence s'est tenue au moment où paraissait refaire surface un vieux problème, avec une urgence renouvelée: comment et par l'intermédiaire de quelles tribunes et quels mécanismes publics parvient-on à l'évaluation et à la compréhension de la croissance globale et des niveaux des dépenses gouvernementales d'une part, et de l'efficacité et du rendement de certains programmes et certaines activités spécifiques du gouvernement d'autre part. Au cours de la seconde moitié des années 1970, le gouvernement a fait l'objet de critiques et de blâmes comme jamais auparavant; la conférence a traduit, en bonne partie, ce problème. Elle a reflété en outre le point de vue fort légitime selon lequel le changement est nécessaire et réalisable.

Toutefois, les conceptions du changement dénotaient une sorte de sobriété intellectuelle et pratique pourtant absente de la précédente vague de réforme de la méthodologie et de l'évaluation, au cours des années 60, d'où devaient surgir le PPB, la gestion par objectifs, l'analyse des systèmes et autres méthodes semblables. Le ton de la conférence indique que le réformateur de l'évaluation des années 80 sera accueilli à la fois comme un missionnaire et un vendeur d'autos usagées. Cette saine ambivalence s'est trouvée renforcée par les questions que sous-tendait l'autre dimension de la conférence, soit les diverses définitions de l'évaluation collective et les nombreuses tribunes où elle peut s'exercer. Pour certains, "collective" signifiait simplement plus d'ouverture dans les affaires gouvernementales, alors que pour d'autres, le sens s'étendait plus largement à la population canadienne et au citoyen votant, homme ou femme, à différents titres. Certains commentateurs, tels M. Pierre-Paul Proulx, ont préconisé le recours à des tribunes et à des publics régionaux décentralisés. Les diverses tribunes d'évaluation collective englobent des scènes aussi diversifiées que celles du Parlement, des média, des groupes d'intérêt, des universités et d'autres instances gouvernementales du fédéralisme canadien.

Chacun des documents et des commentaires de cet ouvrage reflète cette prudente ambivalence concernant la réforme et l'évaluation des dépenses. Cependant, dans ce chapitre d'introduction, nous mettrons en relief quatre thèmes que la conférence a fait ressortir et qui nous paraissent particulièrement importants pour la compréhension des perspectives et des dimensions à long terme de l'évaluation des dépenses publiques. Nous nous pencherons tout d'abord sur le glissement marqué des préoccupations qui, pendant les années 60, s'attardaient à la méthodologie, alors qu'on s'interroge maintenant sur les motifs du discrédit de ces méthodes et sur le peu de changement qu'elles ont entraîné. La seconde partie de ce chapitre portera sur les acteurs motivés par l'intérêt personnel et sur les stimulants et les obstacles qu'ils rencontrent dans le processus d'évaluation, au sein de plusieurs des principales tribunes et institutions telles l'exécutif, le Parlement et les média. Nous tenterons de démontrer, dans la troisième section, comment la perspective de réforme repose sur la compréhension de ces stimulants, la conception de nouvelles tribunes publiques, spécialement d'origine non gouvernementale. Enfin, nous tenterons d'apprécier jusqu'à quel point la vague de réforme plus récente sert le juste équilibre démocratique entre la faculté des gouvernements d'administrer (c'est-à-dire de mobiliser le pouvoir politique) et la nécessité de contenir les abus de pouvoir ainsi que de susciter un sentiment salutaire de précarité chez les élus du peuple. Cet équilibre doit être atteint, particulièrement dans le cadre d'une évolution qui tend à prouver qu'il faudrait bien souvent plus de temps à l'administration pour prendre ses décisions.

LA MÉTHODOLOGIE: D'OPTIMISME EN RÉALISME

Chacun des travaux de cet ouvrage témoigne, à sa façon, d'une certaine modération en ce qui a trait aux problèmes et aux perspectives d'évaluation des dépenses. On affiche bien peu d'enthousiasme pour la "formule technique" rapide qui semblait caractériser si souvent les balbutiements de l'analyse "rationnelle" des politiques, dans les années 60. L'analyse de M. **Irwin Gillespie** insiste sur les lacunes des méthodes techniques. Son exposé sur le comportement du votant gladstonien, keynésien et 'ksanien (étiquette ingénieuse, conçue au Canada, désignant le votant "socialiste" dont la préoccupation principale est la redistribution des richesses) montre clairement l'ensemble des normes qui teinteront l'évaluation et son interprétation. On notera la présence de normes particulières à certains types de dépenses, ainsi que l'expression de préférences relatives pour l'efficacité, la stabilité du revenu et la redistribution. En conclusion, Gillespie soutient que le rôle de l'analyste dans l'ensemble du débat sur l'évaluation des dépenses publiques est de caractère technique et limité.

Le document que présentait **Harry Rogers**, nouveau Contrôleur général du Canada, met également l'accent sur l'importance capitale du sens pratique et du réalisme. Il conçoit le rôle de son cabinet comme celui d'un catalyseur "favorisant une administration des dépenses gouvernementales plus efficace, plus compétente et plus réceptive". Il insiste sur le fait que son bureau ne tentera pas d'instaurer "des systèmes gigantesques et coûteux dans l'espoir de faciliter la solution de problèmes complexes et variés". Selon sa propre expérience sur le terrain, dans le secteur privé, à la fin des années 60, les partisans des systèmes ont entrepris une tâche au-dessus de leur force. Les commentaires du Contrôleur général reflètent un souci de recherche d'une définition plus précise du "programme"; il vise en outre à déterminer si tous les programmes sont susceptibles d'évaluation. Sous plusieurs aspects, la stratégie se veut "invitation discrète" à l'évaluation de programme, à une époque de restriction des dépenses.

L'évolution de différents systèmes budgétaires

L'analyse comparée que fait **Aaron Wildavsky** de l'évolution récente de différents systèmes budgétaires et de l'instinct de conservation profondément enraciné du budget traditionnel (et couramment discrédité) section-article apporte un élément de plus en faveur d'une approche réservée. Le conférencier insiste sur les différents buts, les postulats temporels et, par ricochet, les situations de conflit entre les divers systèmes de contrôle budgétaire. Avant tout, il situe son analyse dans le contexte du comportement relié à la perception des budgets, tout spécialement celui des bureaucraties gouvernementales. Les évaluateurs de programmes de l'avenir devront faire face, par exemple, à l'affirmation logique suivante de Wildavsky au sujet de l'évaluation par programme:

Prenons le cas de celui qui doit décider de l'achat d'une cravate ou d'un mouchoir. Fort simple, direz-vous. Mais supposons que les règles d'organisation exigent l'harmonie parfaite de la garde-robe. La probabilité d'une quelconque modification sera faible si elle entraîne un rajustement complet. Plus le lien entre les différents articles sera étroit et plus les articles seront originaux, plus grand sera le risque d'erreur (la marge de tolérance étant d'autant plus réduite), et plus faible la probabilité de rectification (puisque s'il y a changement, chaque article devra être rajusté avec tous ceux qui l'ont été au préalable). L'alternative de la révolution (c'est-à-dire du bouleversement total) ou de la résignation (soit l'absence de changement) a certes bien peu d'attrait.

Le coût de redressement d'une erreur se trouve augmenté, et non réduit, par l'établissement de budgets par programme. Leur rigidité est le principal reproche qu'on adresse aux bureaucraties. Les choses sont telles que la faveur de l'organisation va au bureau qu'alimentent les catégories usuelles section-article desquelles procèdent les ressources humaines, monétaires et matérielles. Dans l'optique du bureau, les programmes sont négociables jusqu'à un certain point; certains d'entre eux seront diminués et d'autres majorés, l'équilibre de l'agence étant maintenu ou, si nécessaire, ajusté aux périodes de restriction, sans que son existence soit remise en question pour autant. Le budget section-article est plus flexible, précisément parce que les catégories en cause (personnel, entretien, approvisionnement) n'ont pas de lien direct avec les programmes. L'abandon des objectifs, dans un budget par programme, menace l'existence même de l'organisation qui en tire ses fonds, étant donné que le flux d'argent est cette fois orienté vers les objectifs. Il vaut donc mieux que les catégories budgétaires normalisées soient des rubriques sans rapport avec le programme, de sorte qu'il s'offre une diversité de perspectives analytiques; on évite ainsi de recourir à un procédé analytique temporaire comme instrument permanent de répartition des fonds.

Une bonne organisation aura soin de déceler et de corriger ses propres erreurs; plus l'erreur sera onéreuse—non seulement du point de vue monétaire, mais aussi en termes de personnel, de programmes et de prérogatives—, moins il y aura de chances qu'on s'en préoccupe. Par conséquent, les organisations devraient être structurées de telle sorte que les erreurs deviennent repérables et rectifiables, c'est-à-dire qu'elles deviennent évidentes et réversibles, bref que la rectification soit économique et réalisable.[1]

Selon Wildavsky, il ne faut jamais perdre de vue les objectifs, mais il ne paraît pas toujours logique d'établir budgets et évaluations en fonction de ces objectifs, surtout pas si l'on s'intéresse à la réalisation des changements. Ce point de vue deWildavsky est souvent accueilli avec grand étonnement par les défenseurs d'opinions fort abstraites sur le processus rationnel de prise de décisions qui ne parviennent pas à saisir la nature et la logique du comportement politique. Les objectifs politiques et ceux des programmes impliquent non seulement des schèmes de valeurs différents chez divers groupes et bénéficiaires, à un moment donné, mais encore des changements dans le temps, même dans l'intervalle de très brèves périodes. Le conflit politique se loge souvent à l'enseigne des bons.

---

[1] Voir les pages 68-69 de ce volume.

Les programmes et les objectifs

L'analyse des politiques démontre la persistance de leur incohérence. Les ministères doivent souvent obéir à des directives politiques et des dispositions statutaires concurrentes. Plusieurs programmes sont appuyés à des fins différentes, parfois même incompatibles. Il arrive souvent que, pour une large part, la prétendue "administration" ou la "mise en oeuvre" d'un programme ne soit rien d'autre qu'une arène de plus où sont réexaminées les divergences politiques antécédentes. Les individus et les groupes qui ont "perdu la partie" au stade préliminaire d'évolution d'une politique reprennent le combat. Le ton du dialogue politique change souvent. Les divergences s'expriment alors en ces termes: "nous n'avons pas le personnel nécessaire", ou bien "la réglementation n'est pas juste", ou encore "il nous faudra beaucoup plus de temps", tandis qu'antérieurement, les désaccords se fondaient sur les objectifs et les valeurs réels. Les évaluateurs de dépenses devront affronter une réalité de la vie politique, celle du manque de cohérence des politiques. De plus, les groupes et institutions dont les visées sur un programme sont manifestes voudront toujours savoir autour de quels intérêts se poursuit l'évaluation.

LES STIMULANTS ET LES OBSTACLES RELIÉS À L'ÉVALUATION

Les travaux et discussions de la conférence ont mis en évidence la reconnaissance croissante de la motivation et des stimulants, dictés par l'intérêt personnel, qui affectent radicalement le comportement des institutions et des intéressés dans le processus de responsabilité publique. La structure des mesures incitatives de plusieurs tribunes n'est tout simplement pas favorable à la poursuite et à l'usage du type d'évaluation des dépenses que plusieurs réformateurs envisagent. En vérité, les obstacles à l'évaluation sont érigés en système dominant.

Le rôle des média

L'exposé de **Dian Cohen** sur le rôle des média (tout spécialement de la presse) dans l'évaluation des dépenses publiques, quoique anecdotique, témoigne de l'attention passagère de la presse, de la pression qu'impose l'heure de tombée hâtive, de la rotation des affectations professionnelles et souvent de l'insuffisance des connaissances; tous ces facteurs découragent l'évaluation continue des dépenses publiques. Mlle Cohen établit, de plus, qu'un journaliste devenu particulièrement compétent en ces matières reçoit et accepte fréquemment l'invitation d'un ministre à se joindre à son personnel politique. Au mieux, seulement une poignée de journalistes canadiens possèdent la compétence et manifestent suffisamment d'intérêt pour consacrer un peu de leur temps à l'évaluation des dépenses, entre autres de certains programmes spécifiques.

L'absence de stimulants

Michael English a procédé à l'analyse des méthodes d'examen des dépenses publiques au Royaume-Uni; au cours de la conférence, les commentaires de **James Gillies**, député, et **Maurice Leclair**, secrétaire du Conseil du Trésor, viennent confirmer les analyses récentes de Dobell[2] et Thomas[3] qui démontrent le manque de mesures favorisant la poursuite et l'usage de l'évaluation des dépenses par les institutions parlementaires. C'est le cas, tout spécialement, de la Chambre des communes du Canada où les normes de politique partisane s'affirment fortement, de même que la discipline de parti et la conviction que le Parlement est le théâtre de combats de gladiateurs; il ne peut s'ensuivre au mieux qu'un intérêt très sélectif en matière d'évaluation. Les structures incitatives se retrouvent, dans l'ensemble, au centre de l'arène, là où le processus de responsabilité parlementaire est soumis aux règles strictes du besoin de confiance et de la discipline du parti.

Le pouvoir exécutif

La prédominance du pouvoir exécutif que renforcent les lois et traditions du Cabinet, de même que le secret administratif d'une part et, d'autre part, les politiques d'information et les fuites calculées, font la faiblesse des comités parlementaires et autres mécanismes d'enquête. Tout cela trouve appui et encouragement dans la dominance historique d'un parti politique sur la scène fédérale et dans l'ambivalence que manifestent stratèges et leaders de l'opposition à l'égard du rôle des partis d'opposition. Cette ambivalence oscille entre "l'opposition pour l'opposition", "l'opposition à caractère constructif" et l'opposition à titre de "gouvernement éventuel". De plus, comme le note James Gillies, la stratégie de l'opposition pendant la période de questions a fortement tendance à subir l'influence marquée des reportages quotidiens de la presse et des autres média. Donc, les stimulants offerts par les média viennent renforcer ceux de l'arène parlementaire.

Il faut établir que toutes les mesures incitatives du Parlement ne s'érigent pas contre l'évaluation des dépenses. La lutte parlementaire se transforme souvent en tribune d'évaluation et d'information politique traditionnelle où les points de vue de rechange sur les niveaux de dépenses et les normes de programmes sont affirmés et viennent conditionner les choix du gouvernement, même s'il s'agit souvent de *prévenir* la critique parlementaire. De plus, notons des exemples d'évaluation très efficace des programmes de la part du Comité sénatorial des finances, notamment en ce qui a trait aux programmes et actions touchant la politique de main-d'oeuvre,

---

[2] Peter Dobell, 1977.

[3] Paul Thomas, 1977.

Information Canada et le programme de logement du ministère des Travaux publics.

La présence d'obstacles

L'existence d'obstacles à l'évaluation significative chez l'exécutif et la bureaucratie devient de plus en plus évidente. L'analyse comparative de Wildavsky jette une abondante lumière sur les obstacles attribuables à la bureaucratie. La structure incitative bureaucratique attache une importance disproportionnée aux dépenses, à la proposition de nouveaux programmes et aux bénéfices à relativement court terme. Son influence négative se remarque à l'égard des programmes et de la planification à long terme et de la préparation de l'information préalable à l'évaluation. En outre, quand les renseignements existent déjà, l'incitation à déformer ou à soustraire ces renseignements aux ''non-initiés'' qui pourraient les utiliser au désavantage d'une agence apparaît clairement; dans le contexte bureaucratique, l'information est une monnaie essentielle au pouvoir.

L'analyse de **Michael Prince** concernant les organismes d'évaluation du gouvernement fédéral au Canada présente d'autres facettes du dilemme.

> On comprendra sans peine pourquoi il y a si peu d'agences ou de ressources des ministères consacrées à l'évaluation de programmes déjà établis. Quand des groupes de conseillers en politiques évaluent les programmes gouvernementaux, ils évaluent la raison d'être des ministères qui tirent de ces programmes leur appui, leur force et leur justification, à l'instar des agences. Donc, une évaluation complète et investigatrice pourrait être interprétée comme une menace pour le personnel des programmes. Les hauts fonctionnaires et le crédit qu'ils accordent à l'analyse, la nature d'une politique ou d'un programme, le poids politique que représente la clientèle de ce programme, l'usage que le ministère a fait, dans le passé, des méthodes d'analyse et d'évaluation jouent sur l'attitude qu'adopte ce dernier à l'égard des bienfaits de l'analyse des politiques.
>
> Lorsque des groupes sont munis de ressources destinées à l'évaluation des programmes, plusieurs conseillers en politiques estiment que leur rôle d'évaluateur entre en conflit avec les autres fonctions qui leur sont assignées, de même qu'avec la nécessité d'établir de bons rapports avec les fonctionnaires organiques du ministère, surtout au cours des premières années d'application du programme. Par conséquent, les conseillers ont prôné des initiatives plus faciles sur le plan technique et moins menaçantes du point de vue de l'organisation, telles l'élaboration de politiques, au détriment de l'évaluation des programmes. Chez quelques groupes, le rôle de brigade d'incendie a prévalu, consciemment ou non, empêchant, en certains cas, l'évolution des rôles de planification et d'évaluation. Il ne faut pas croire que le problème sera résolu simplement en plaçant les groupes à l'abri de la fièvre des soucis quotidiens.
>
> D'autres contraintes et considérations gênent les groupes dans l'organisation de leur rôle de conseillers, dont l'insuffisance des ressources (temps, personnel, classification des tâches et support matériel), les difficultés de recruter du personnel qualifié et de le retenir en service, l'inexpérience de quelques conseillers, l'absence de données pertinentes, ainsi que des problèmes techniques dans l'application des méthodes analytiques. De même, les caractéristiques distinctives des conseillers en politiques ont affecté les rapports hiérarchico-fonctionnels et le rendement du groupe. Sans doute une réaction négative s'est-elle manifestée chez certains administrateurs et certains fonctionnaires des ministères à l'égard des groupes de conseillers en

politiques composés de "génies en herbe". De plus, le rôle de ces groupes était souvent imprécis ou rejeté par les fonctionnaires organiques. Ainsi privés d'appui et éloignés des réseaux d'influence et de persuasion, quelques groupes sont demeurés à l'écart des grands courants de la politique ministérielle et des processus de gestion.[4]

Les stimulants du milieu universitaire

En ce qui a trait aux stimulants du milieu universitaire, l'insistance de Irwin Gillespie sur les limites de la technique méthodologique et sur la primauté des normes et valeurs en concurrence ne rallie probablement pas la majorité des opinions du milieu. Le système de stimulants y fait encore l'éloge de la technique d'évaluation et de la recherche d'une connaissance causale. L'importance croissante des contrats de recherche et de consultation qui s'offrent aux universitaires intéressés et versés dans l'évaluation des politiques et des programmes peut susciter l'absence d'indépendance, à la fois en apparence et en substance. A un autre palier, l'intervention du milieu universitaire se révèle d'intérêt limité dans le domaine de l'évaluation collective en raison de la rigidité de la discipline du milieu et du fait que la communication a tendance à s'établir davantage par le truchement de journaux d'étiquette huppée et de tutelle savante que par celui d'autres catégories de média et de publications.

Comme en fera foi, nous l'espérons, la section suivante de ce chapitre d'introduction, nous n'avons pas l'intention, par la précédente critique des systèmes de stimulants, de soutenir naïvement que chacune de ces institutions devrait se transformer totalement. Les systèmes incitatifs dominants dans chaque arène ou chaque tribune se perpétuent dans le temps parce qu'ils servent également à remplir d'autres fonctions intentionnelles et valables qui échoient à ces institutions. Si toutefois l'on n'a pas conscience que ces stimulants peuvent avoir des conséquences néfastes sur l'évaluation collective, des propositions de réforme irraisonnées surgiront, susceptibles d'entraîner directement l'élaboration de politiques et de choix collectifs au moins partiellement sous l'influence de stratégies ayant pour objet d'éviter l'évaluation directe. Plusieurs des travaux de cet ouvrage insistent sur ce point, exemples à l'appui.

Les dépenses fiscales

L'exposé de M. **Allan Maslove** sur les dépenses fiscales montre comment un instrument de politiques relativement occulte jusqu'ici—soit l'utilisation du système fiscal pour dispenser des bénéfices—serait de plus en plus privilégié par les instances de décision, au détriment d'autres moyens directs, plus visibles (tels que les dépenses fixes), le recours à la fiscalité étant précisément moins susceptible d'évaluation et ces choix étant plus

---

[4] Michael Prince, 1979.

discrets dans la documentation publique. L'évaluation initiale que fait Maslove des dépenses fiscales au Canada démontre leur orientation régressive prononcée; il recommande que le Ministre des Finances soit tenu de publier les données relatives aux bénéfices, ce qu'exige désormais la loi américaine.

La croissance de la bureaucratie

Dans une veine quelque peu différente, MM. **Richard Bird** et **David Foot** brossent le tableau de la croissance bureaucratique. Leur étude fait état de la nécessité de l'évaluation empirique et des difficultés qu'elle comporte. Le milieu des affaires et des grandes sociétés formule machinalement bien des critiques concernant l'accroissement des dépenses publiques, critiques qui découlent implicitement ou explicitement de croyances et d'assertions sur la croissance de la bureaucratie et la soif d'expansion des bureaucrates, comme l'indiquent les commentaires de M. **William Twaits**, ex-Président de la Compagnie pétrolière impériale, lors de la conférence. MM. Bird et Foot se gardent bien, dans leur texte, de toute allusion à un lien simple de cause à effet entre l'accroissement de la bureaucratie et celui des dépenses; ils n'en évoquent pas moins un certain nombre de mythes concernant les paramètres de la croissance de la bureaucratie. Cette analyse souligne la différenciation nécessaire entre "l'emploi dans le secteur public" et "la fonction publique" et attire l'attention sur le taux de croissance relativement rapide de l'embauche dans le secteur public au cours des premières années d'après-guerre.

Les analyses des dépenses fiscales et de l'accroissement bureaucratique font voir la nécessité et la valeur d'une prudente évaluation collective. Sans nul doute, pareilles "analyses" passeront par divers tamis idéologiques. Les croyances et postulats idéologiques peuvent se substituer à d'autres efforts de réflexion et devenir l'expression fruste d'un ensemble de valeurs. Ce qu'on dit des fins présumées de l'idéologie sur la place publique se révèle grandement exagéré.[5] L'idéologie ne détermine pas nécessairement les politiques, mais elle contribue presque toujours à rétrécir le *champ* des options considérées et influe sur la façon d'accueillir et de traiter les évaluations formelles. De plus, des domaines de programmes spécifiques s'avéreraient tout spécialement sensibles à l'influence des deux grandes idéologies (celles de gauche et de droite), ainsi qu'à des paradigmes ou idées quelque peu plus étroits, mais tout de même prégnants (la valeur que l'on accorde à l'éthique du travail dans plusieurs programmes sociaux, par exemple).[6] Ainsi peut-on prétendre, en réaction à l'évaluation des dépenses fiscales, qu'il vaut mieux ne pas les définir comme des taxes que le

---

[5] Daniel Bell, 1970.

[6] R. Manser, 1975 et Peter Aucoin, 1979.

gouvernement a décidé de ne pas percevoir, toutes les ressources se trouvant, dans cette optique, propriété de l'État, mais plutôt les considérer comme des taxes que le gouvernement n'a pas lieu de réclamer. De même, des évaluations subtiles de l'accroissement de la bureaucratie ne sauront pas convaincre quiconque croit tout simplement à une "trop forte" croissance gouvernementale.

Le commentaire de M. **Hugh Armstrong** met l'accent sur les fondements politiques inhérents à l'évaluation. Celle du secteur public est, par nature, sélective et se déroule à travers des *inégalités* patentes, tant dans les occasions offertes à diverses communautés d'évaluer et d'attirer l'attention des autorités politiques que dans la distribution réelle des bénéfices de l'activité gouvernementale (dans les domaines de la réglementation et de la fiscalité). Par conséquent, aux yeux de M. Armstrong, il ne faut pas s'étonner de ce que les gouvernements ne s'attardent pas tout d'abord à l'évaluation des conséquences des programmes sur la redistribution.

Le ton général des communications à la conférence démontre clairement qu'en dépit de l'existence de quelques problèmes techniques d'évaluation, la question des stimulants et de l'intérêt se situe au coeur des stratégies de réforme à venir.

LA RÉFORME ET LA CONCEPTION DE STIMULANTS NOUVEAUX

Rien ne saurait être aussi évident que l'absence de remèdes simples et d'un seul résultat "correct" dans le processus d'évaluation des dépenses publiques. La tâche principale demeure donc la conception et la mise en oeuvre de stimulants nouveaux qui, dans l'ensemble, susciteront des modèles de comportement nouveaux ou différents grâce auxquels l'évaluation ouverte ne sera pas perçue comme une menace par ceux qui en font l'objet; par conséquent, elle deviendrait moteur de changement là où il s'impose. Reconnaissons également l'importance d'édifier sur la base des progrès accomplis dans le sillage de la vigoureuse réforme analytique des années 60. Comme **Dean Crowther** l'évoque dans son examen de l'expérience américaine, la période de réforme antérieure a contribué à la production de "meilleurs chiffres" et a entraîné en plusieurs milieux la reconnaissance, même réticente, du besoin d'analyse. Dans le contexte canadien, la réforme future peut s'édifier sur les phases progressives antérieures de l'industrie de "l'analyse des politiques". Cette industrie a pris corps au milieu des années 60, avec la création d'agences centrales de conseillers en politiques, telles que le Conseil des sciences et le Conseil économique, et leurs premiers rapports. Bientôt, vers la fin des années 60, apparaissaient de petits organes de "planification" au sein des agences centrales. Les fonctionnaires de ces agences, à leur tour, encourageaient la formation de groupes de conseillers en politiques dans les ministères; ces groupes ont proliféré rapidement au début des années 1970. Les fonctionnaires des agences centrales étaient le fer de lance de l'expérience qui a conduit à la création de ministères consacrés à l'élaboration de politiques, comme le ministère d'État aux Sciences et à la Technologie et le ministère d'État chargé des Affaires urbaines, dont la sphère d'influence au départ devait être le pouvoir d'acquisition de l'information et d'analyse. L'industrie de l'analyse des politiques a aussi pris une saveur autre que gouvernementale avec l'émergence et le renforcement de plusieurs "laboratoires de réflexion" sur les politiques et groupes indépendants ou du secteur privé, comme l'Institut de recherche C.D. Howe, l'Institut de recherches politiques, le Conference Board au Canada, l'Institut Fraser, la Fondation Canada West et le Conseil économique de l'Ontario, au milieu des années 70.

Doern et Aucoin soumettent les remarques suivantes:

> Les propositions en vue d'instituer des systèmes d'évaluation de programmes dans tous les ministères fédéraux se voulaient en partie une réaction contre l'évolution de l'industrie de l'analyse des politiques et, en partie également, la poursuite de son action. Une réaction négative en ce sens que le mouvement d'évaluation de programmes avait pour but de façon plus pragmatique d'assurer que les programmes courants s'appliquaient effectivement et efficacement. Ce pragmatisme a reçu l'encouragement du Vérificateur général qui exhortait à obtenir "plus pour son argent" et celui d'un point de vue de plus en plus répandu selon lequel les efforts antérieurs plus grands déployés pour l'analyse des politiques avaient fait surgir des groupes presque semblables à des universités sans étudiants. Ce n'est pas sans raison qu'on était sceptique à l'égard de l'évolution d'une version nouvelle de "paralysie par l'analyse" . . . [mais] il en est résulté à la fois des bénéfices et des coûts. Un processus plus concurrentiel de dispensation de conseils sur les politiques a été conçu de telle sorte que de plus larges avenues se sont ouvertes, favorisant les échanges et l'émulation entre les agences centrales et les ministères, entre les fonctionnaires d'état-major et les fonctionnaires organiques. Plusieurs personnes ayant d'excellentes idées ont ainsi fait leur entrée dans les cercles gouvernementaux.[7]

La nécessité de lignes de pensée concrètes concernant les stimulants se fait peut-être plus pressante à l'examen de l'avalanche de propositions de réforme qui ont été formulées au cours des dernières années. Citons entre autres:

a) la préparation et la publication d'un Livre blanc sur les dépenses publiques qui présenterait les projections et les schémas de dépenses gouvernementales, à moyen terme (de trois à cinq ans);
b) l'adoption de lois sur la dissolution des organismes en vertu desquelles les organismes et programmes disparaîtraient automatiquement à moins d'être rétablis à la suite de l'évaluation de leur rendement;
c) la préparation de budgets à base zéro et l'instauration d'une évaluation interne des programmes, systématique et continue; et
d) l'adoption de lois sur l'information du public et autres mesures réformatrices qui donneraient accès à l'information que produit et

---

[7] Doern et Aucoin, 1979.

monopolise la bureaucratie; ces renseignements seraient ainsi davantage accessibles au public et aux autres institutions gouvernementales.

Le Livre blanc sur les projections de dépenses

Plusieurs groupes et individus ont suggéré que le gouvernement fédéral soit tenu de publier et de soumettre au Parlement des projections de dépenses de trois à cinq ans à l'avance.[8] A l'image, en partie, de la pratique britannique, la préparation d'un Livre blanc sur les dépenses publiques au Canada pourrait comporter les étapes suivantes:

— la publication par le Président du Conseil du Trésor d'un Livre blanc annuel sur les dépenses du gouvernement;
— la création d'un Comité parlementaire central responsable du budget de dépenses et de l'économie, dans le but de recevoir et d'analyser le Livre blanc et de solliciter mémoires et témoignages à ce sujet;
— l'organisation d'un débat parlementaire intégral sur le Livre blanc et sur l'évaluation qu'en aura fait le Comité des dépenses.

La préparation du Livre blanc serait confiée au Conseil du Trésor, qui en assumerait également la publication, et au ministère des Finances. Les prévisions internes de programmes, déjà effectuées par le Conseil du Trésor, sans être rendues publiques, seraient l'un des éléments de base du Livre blanc.

Ce Livre blanc contiendrait des renseignements sur les schémas et les projections de dépenses pour une période de trois ans, en fonction des programmes et des ministères et selon les catégories économiques de dépenses. Ce document fournirait également la projection des revenus annuels pour la période de trois ans, ainsi que des données sur les prévisions économiques à moyen terme du gouvernement et les hypothèses sous-jacentes.

Le Président du Conseil du Trésor soumettrait le Livre blanc au Parlement à la fin de l'année civile, avant la présentation en Chambre des prévisions budgétaires annuelles. Cela pourrait donner lieu éventuellement à un débat parlementaire de deux jours, faisant suite à l'évaluation initiale remise à la Chambre des communes et aux commentaires du Comité responsable du budget de dépenses et de l'économie, tel que proposé.

Outre la préparation d'un commentaire initial sur le Livre blanc, ce comité pourrait produire d'autres rapports et recueillir des témoignages publics de divers groupes intéressés et de groupements d'experts, tels que le Conseil économique du Canada, et d'organismes privés, comme, entre autres, l'Institut de recherche C.D. Howe. Un tel comité devrait compter sur l'appui d'un personnel adéquat pour bien s'acquitter de ces responsabilités.

En ce qui a trait à l'investigation et à la responsabilité parlementaires, on

---

[8] G. Bruce Doern, 1978 et Hal Kroeker, 1978.

affirme que pareil processus susciterait l'apparition d'un forum où se créerait une réaction collective d'opposition au total des dépenses et, par définition, prolongerait la durée de l'analyse. L'influence du Parlement s'exercerait vraiment où elle le peut avec le plus d'efficacité, c'est-à-dire dans le champ des priorités de dépenses *à venir*, à moyen terme. Il semble évident que son influence sur les dépenses de l'année courante est faible ou même nulle. Du point de vue du processus *interne* de responsabilité au pallier de l'exécutif, on a fait valoir que le procédé du Livre blanc contribuerait à faire tomber les barrières entre les ministères; contrairement à la pratique courante des agences centrales qui gardent pour elles les prévisions de programmes et autres données concernant les dépenses, les ministères connaîtraient mieux et plus ouvertement leur position exacte, les bases de projections et de données étant connues.[9]

Pour bien comprendre comment pourrait advenir une telle réforme des dépenses, on doit se référer à l'évolution et la politique des dépenses et des budgets gouvernementaux, comme Aaron Wildavsky y faisait allusion.

Les attitudes politiques et les dépenses gouvernementales

N'est-ce pas vérité évidente que les attitudes politiques à la mode, à l'égard des dépenses gouvernementales, exercent une grande influence sur la gestion financière et la responsabilité du gouvernement. Au Canada, comme en d'autres pays occidentaux, on a noté trois phases distinctes d'attitude et de démarche vis-à-vis des dépenses gouvernementales. Du milieu des années 30 jusque vers la fin de la décennie de 1950, malgré la croissance accélérée en nombre et en portée des fonctions gouvernementales, le souci de frugalité et de régularité dans les dépenses et la gestion financière a dominé. Le rythme annuel de l'activité budgétaire était centré sur les prévisions annuelles (réunies en catégories courantes de dépenses, telles que les salaires et les approvisionnements).

John Maynard Keynes a marqué de son influence dominante les années 1950 et particulièrement la décennie 1960. On en est venu à considérer les dépenses gouvernementales comme outil de gestion de la demande dans l'élaboration des politiques fiscales, influant ainsi sur les attitudes à l'égard des dépenses de l'État. Donc, au souci initial de frugalité et de régularité s'ajoutaient les exigences de l'économie keynésienne.

Un troisième ensemble d'attitudes à l'égard du processus de dépenses apparaît avec le début de la décennie de 1960, époque de la publication au Canada du rapport Glassco et des rapports Plowden et Fulton, au Royaume-Uni. Alors que les programmes de dépenses gouvernementales connaissaient une augmentation rapide, on s'est surtout soucié de la gestion positive des ‘‘programmes’’. Cette phase a été également caractérisée par des

---

[9] G. Bruce Doern, 1977.

efforts pour étendre à une période de trois à cinq ans le cadre temporel de la planification des dépenses, à l'intérieur du gouvernement, qui se bornait auparavant à des prévisions annuelles. Les prévisions budgétaires étaient catégorisées selon la combinaison des programmes individuels axés sur les extrants, reliés aux objectifs de programmes de même qu'aux catégories courantes traditionnelles de dépenses.

Le processus de dépenses (et, en définitive, la gestion financière) subissait l'influence successive de trois sortes de valeurs ou de résultantes. La présence de chacune d'elles s'expliquait en soi, mais leur réunion avait tendance à semer la confusion et à distraire l'attention des ministres, des hauts fonctionnaires ainsi que des principaux groupes économiques. Toute réforme éventuelle du processus de dépenses, dont le projet de Livre blanc sur les dépenses et son examen, devra tenir compte des valeurs souvent incompatibles que doivent concilier les dépenses publiques.

Les degrés de contrôle

Des conflits de temps aussi bien que de valeurs surgissent à propos des différents *degrés* de contrôle des dépenses publiques de toutes sortes. Ce contrôle sera plus ou moins efficace en raison des différentes juridictions politiques (les gouvernements provinciaux au Canada, par exemple) ou variera selon le type d'agence gouvernementale ou d'entité (ex., les sociétés de la Couronne). Le contrôle peut différer selon qu'il s'agit d'immobilisations ou de dépenses courantes, de dépenses non discrétionnaires en réponse à la demande (ou statutaires et inévitables), de dépenses dont l'impact sur les produits et le marché du travail varie (les coûts de construction, par exemple) ou de dépenses qui se prêtent à la gestion de la demande à court terme, par opposition à celles qui ne s'y prêtent pas.

L'expérience britannique, comme d'ailleurs celle de la politique et du gouvernement canadiens, démontre nettement que la mise en oeuvre d'un processus comprenant la publication et l'examen d'un Libre blanc sur les dépenses gouvernementales n'ira pas sans difficulté, tout au moins au départ et peut-être à jamais. Comme en témoigne l'expérience britannique, des résistances se manifesteront à la fois dans les sphères de l'exécutif et du Parlement, pour les mêmes motifs que là-bas. Il appert également qu'un gouvernement fédéral pourrait s'opposer catégoriquement à pareille proposition, si les administrations provinciales n'étaient pas uniformément prêtes à se compromettre par la publication de leurs plans et projections de dépenses. Selon toute probabilité, le débat public ainsi engagé se poursuivrait entre experts, puisque seulement une poignée de journalistes financiers et d'universitaires manifesteraient assez d'intérêt et consacreraient assez de temps pour y participer.

Toute réforme obtient l'appui ou suscite la désapprobation d'une variété d'intéressés dont les motifs sont fort divergents. Certains considéreront le

Livre blanc comme un moyen ingénieux d'instituer et de perpétuer un conservatisme politique et fiscal, en exigeant du gouvernement qu'il établisse un rapport entre revenus et dépenses, de façon mieux intégrée. Toutefois, cela pourrait à tout le moins signifier un "élément de complication" pour le gouvernement qui se verrait confronté chaque année avec les écarts inexpliqués entre les projections datant de trois ans et le niveau des dépenses courantes. D'autres intéressés verraient dans cette réforme un mécanisme utile à la compréhension et l'explication de la tendance des dépenses. Ces données, divulguées au public, indiqueraient quels postes ont besoin de fonds et pourquoi, dans des programmes spécifiques de portée sociale et, par conséquent, l'on verrait avec plus de clarté et de régularité les gagnants et les perdants du processus de "redistribution" des ressources dont nous avons déjà parlé.

Les lois sur la dissolution des organismes

Des lois sur la dissolution des organismes ont été adoptées dans plusieurs États américains et ont été proposées au Canada.[10] La loi sur la dissolution est plutôt considérée comme instrument de réforme des organismes de réglementation, mais, en fait, elle pourrait convenir à tout ministère, toute agence gouvernementale ou société de la Couronne. Le parti conservateur de l'Ontario proposait une telle loi en mai 1976, lors de son congrès annuel. A son entrée en vigueur, elle exigerait des commissions de contrôle provinciales qu'elles justifient leur existence tous les deux ans sous peine de dissolution. Les partisans de la proposition ont soutenu que, par ce moyen, l'intervention gouvernementale diminuerait de même que la croissance des bureaucraties publiques. Un concept similaire était suggéré dans un énoncé de politiques publié par le parti progressiste-conservateur fédéral, en 1978.[11]

Les lois sur la dissolution des organismes semblent destinées à conférer plus de transparence au phénomène de la réglementation. La réglementation est une dimension plutôt occulte du gouvernement du fait qu'agrégée ou marginale, elle n'a pas été aisément convertie au dénominateur commun des dépenses, tout spécialement dans ses effets sur les budgets privés (individuels et corporatifs). Le mouvement d'identification des "dépenses fiscales" se conçoit comme un effort pour mettre ces choix en évidence. Les lois sur la dissolution des organismes, cependant, semblent être issues d'hypothèses héroïques sur la possibilité du Parlement d'évaluer organismes et programmes, à court terme, à même la masse d'un feuilleton parlementaire chargé. Dans ce contexte, l'expérience de révision de la *Loi sur les banques*, tous les dix ans, est révélatrice.

---

[10] Robert D. Behn, 1977.

[11] *Globe and Mail*, 17 mai 1976 et Parti progressiste conservateur du Canada, 1978.

Le budget à base zéro et l'évaluation des programmes

Des inquiétudes et des conflits de même nature que ceux évoqués ci-dessus sous-tendent les propositions d'adoption de la méthode à base zéro de préparation des budgets et d'évaluation des programmes. Les partisans de la méthode à base zéro envisagent un procédé par lequel *toutes* les dépenses (à l'échelle du Canada, les budgets A et B) seraient évaluées chaque année, de sorte que de plus en plus d'argent ''frais'' provienne de l'annulation ou de la modification d'activités déclarées inefficaces ou inutiles.[12] L'analyse de Wildavsky ne devrait laisser planer aucune illusion dans notre esprit quant au sort prévisible de cette réforme.

Les lois sur la liberté d'information

La réforme apparaît essentielle dans un autre domaine que la conférence abordait toutefois indirectement. Elle implique un important renforcement des lois sur la liberté d'information. Ces lois ne sont pas en elles-mêmes des panacées puisqu'elles peuvent interpréter à leur gré ce qu'est une surcharge d'information; mais il n'en reste pas moins que le système de gouvernement du Cabinet canadien, dans notre régime parlementaire, est excessivement secret. La plupart de nos concitoyens sont désormais d'un enthousiasme modéré vis-à-vis la plus grande liberté d'accès à l'information bureaucratique, mais peu semblent se rendre compte des modifications potentielles du système de stimulants qu'entraîneraient ces lois. Plus que toute autre réforme discutée dans ce volume ou ailleurs, l'accès du public à l'information peut altérer la nature des stimulants bureaucratiques favorisant une évaluation collective plus efficace. La possibilité (la menace) d'une investigation de la part du public exercerait un impact majeur sur la formulation des politiques, leur administration et leur évaluation. Toute tentative sérieuse de réforme dans cette voie trouvera une forte résistance, révélatrice, par sa nature même, de la présence du secret qui, sous bien des aspects, articule le système actuel de stimulants et les modifications potentielles de ces stimulants que l'accès du public à l'information engendrerait.

La question se résume donc non pas aux techniques d'évaluation et à ses résultats, puisqu'il n'y a pas de résultat unique qui soit ''correct'' (comme Gillespie l'a clairement indiqué), mais aux stimulants et procédés d'évaluation. Nous avons souligné auparavant la difficulté que présente la restructuration des stimulants des institutions majeures en vue de susciter une meilleure évaluation collective des dépenses gouvernementales. Il paraît donc essentiel d'adopter des réformes principalement en fonction du potentiel de changement qu'elles peuvent opérer sur le comportement. Les média, le Parlement, les groupes d'intérêt et autres forums publics possèdent tous un système de

---

[12] Peter A. Pyhrr, 1977.

stimulants dont l'évolution s'est échelonnée sur de longues périodes et remplissent aujourd'hui des fonctions importantes et nombreuses, quelques-unes seulement ayant des liens avec l'évaluation des dépenses globales ou spécifiques. Ceci étant dit, il nous apparaît tout spécialement nécessaire que les universités et autres centres de savoir indépendants, tels les instituts de politiques, assument des responsabilités de caractère spécialisé. On serait davantage porté à croire que ces centres de "savoir" prennent leurs responsabilités s'ils manifestaient plus d'habileté et de volonté de communication, plutôt que de s'en tenir à la simple publication de renseignements. Les universitaires sont encore plus enclins à publier pour leurs collègues et à s'adresser à eux qu'à se tourner vers un auditoire plus vaste. La concurrence récente de nouveaux instituts privés de politiques pourrait instaurer de nouvelles habitudes de communication.

Le problème classique de "l'acte gratuit" se pose encore aux évaluations collectives des universitaires et des instituts privés de politiques. Les évaluations collectives constituent un bien public, ce qui suppose qu'on devrait pouvoir les obtenir sans frais. Les conférenciers tels **Carl Beigie**, de l'Institut de recherche C.D. Howe, et **David Slater**, du Conseil économique du Canada, ont soulevé la question suivante: qui devrait assumer la responsabilité d'évaluation des dépenses publiques et en publier les résultats? Ces conférenciers ont mis en évidence la difficulté de se procurer des fonds à ces fins. Par définition, aucun groupe particulier ne tire profit directement des évaluations globales et, par conséquent, chacun affiche une certaine réticence à s'engager dans une initiative d'évaluation du domaine "public".

L'ÉQUILIBRE DÉMOCRATIQUE MODERNE

L'évaluation collective des dépenses gouvernementales repose sur un sain équilibre entre au moins deux types de connaissances, soit le service de renseignements politiques traditionnel—celui des élus du peuple—et la connaissance dont la source réside dans les recherches et les analyses claires, divulguées publiquement. Il est certain que plusieurs institutions ont un rôle à jouer dans l'atteinte de cet équilibre des connaissances. Il s'agit du Parlement, des média, des groupes d'intérêt, des initiatives communautaires et du secteur universitaire. Ils forment une chaîne dont plusieurs maillons sont faibles, au Canada. L'addition d'une "liste de voeux" réformistes ne suffira pas à renforcer la chaîne. Ces réformes doivent être évaluées *collectivement* en relation avec d'autres éléments essentiels de l'équilibre démocratique, celui qui s'établit entre l'assurance de la capacité du gouvernement d'administrer et de mobiliser le pouvoir d'une part, et la présence de contraintes aux abus de pouvoir d'autre part, compte tenu, en outre, de l'impact spécifique des réformes sur les stimulants. L'équilibre des connaissances et celui du pouvoir doivent tous deux être atteints, nécessité d'autant plus impérieuse que des conflits surgissent non seulement au sujet des divers buts que poursuivent les politiques et les programmes, mais également au sujet des échéances imposées aux choix collectifs.

Le conflit des échéances

Plusieurs facteurs ont influé sur le conflit des échéances. D'abord, les réformateurs de politiques d'agences centrales ont suggéré, aspiration légitime, la nécessité d'envisager les choses à plus long terme. D'autre part, le conflit entre les réalités à moyen terme, inhérentes au système de gestion économique de l'offre, et les exigences de "réglage" à court terme de la gestion de la "demande" économique keynésienne est devenu de plus en plus évident. On a reconnu la nécessité explicite d'un sommet économique fédéral-provincial à "moyen terme". Troisièmement, des analyses du processus de réglementation ont révélé une inquiétude sans précédent au sujet de l'ensemble des conséquences à long terme de la réglementation gouvernementale croissante. Pour conclure, des enquêtes du style Berger font également l'objet de demandes, afin de procurer aux individus les plus touchés une évaluation complète des perspectives à long terme des principaux investissements socio-économiques. Doern et Aucoin le soulignent:

> D'une part, la logique commande cette élaboration et cette extension des échéances en matière de politiques. Pourtant, d'autre part, il est compréhensible qu'un État fédéral démocratique libéral répugne à réclamer une économie davantage "planifiée" ou en reconnaître la nécessité. Parallèlement, les demandes à court terme augmentent au point de laisser croire que le gouvernement est débordé et ne peut suffire à la tâche. Compte tenu de la rareté des ressources et de leurs limites, le système politique est contraint soit de provoquer une modération des attentes de la population pour réduire les demandes, soit de satisfaire certaines d'entre elles en rejetant la plupart des autres, ou encore de concevoir un éventail de réponses, selon plusieurs modes de combinaisons et de permutations des comportements suivants: l'expression d'un souci symbolique, la mise en place de forums consultatifs, l'étude du problème (groupe de travail, commission royale, audiences, recours aux "laboratoires de réflexion"), la réorganisation ostensible des agences et des programmes, l'élaboration de nouveaux programmes de dépenses et de réglementation.
>
> La virtuosité croissante avec laquelle on a fait usage de cet éventail de réponses a donné lieu à de nombreuses propositions de réformes vers la fin des années 70.[13]

Chacune des réformes évoquées brièvement dans la section précédente est en partie fondée sur le souci d'évaluation de leurs conséquences à plus long terme. Mais chacune, dans ce contexte d'examen des conséquences à *plus long terme* et d'extension des échéances en matière de politiques, encombre le calendrier des décisions à court terme. Les réformes, certes, favorisent une vision des choses à plus long terme et soulignent le besoin d'approfondir les analyses, mais chacune d'entre elles impose de nouveaux retards et délais à court terme, si bien que l'avenir est plus incertain. La

---

[13] Doern et Aucoin, 1979.

situation finit pas se prêter aux abus. L'analyse et la recherche peuvent trop facilement se substituer aux réactions plus pragmatiques face à des problèmes de politiques et servir des intérêts dont l'objectif serait d'entraver *toute* action gouvernementale. De nombreux motifs justifient l'analyse qu'on invoque pour approuver ou désapprouver plusieurs causes. Les analystes et évaluateurs de politiques ont tout intérêt à promouvoir l'analyse. Ils font partie des compétences du savoir. La concurrence, au plan de l'analyse, entre l'agence centrale et le ministère organique (sans compter la concurrence fédérale-provinciale) peut franchir cette ligne de démarcation invisible au-delà de laquelle l'analyse cesse d'être stimulante du point de vue organisationnel ou politique; elle devient alors improductive et même destructive de l'action concertée. Il est de plus en plus évident que la valeur du jeu de l'évaluation collective des dépenses a des limites. Mais gardons-nous d'être plus zélés que prudents en rejetant l'analyse, car, il n'y a pas si longtemps, rien ne se faisait au Canada, en matière d'évaluation collective des politiques, hormis quelques enquêtes périodiques de commissions royales.

Chapter Two

Fools Gold: The Quest for a Method of Evaluating Government Spending

by
*W. Irwin Gillespie**

> The evaluative stance adopted by the economist in matters of public decision making is derived from our *interpretation of the actual role of the state* as a provider of goods and services vis-à-vis the private sector, plus *our value judgments concerning the proper role of the state* in such matters.
>
> D.W. Bromley

> While the involvement of social scientists as entrepreneurs in social change is not without creating some unease among the liberal-minded intelligentsia, it may turn out to be the only viable and effective strategy while we organize better ways for citizens to participate in social development.
>
> Gilles Paquet

> Cabinet ministers are firmly convinced that economic policy is much too important to be left to economists, or administrators for that matter, just as the conduct of war is too important to be left to generals.
>
> Alan Peacock

PROLOGUE

The analyst who tries to evaluate government spending is embarking upon a quest for fool's gold. He does not recognize it as such when he stumbles upon an evaluative methodology, and is likely to proclaim to all that he has found the real thing and the government is spending wastefully and inefficiently. Politicians and citizens alike ignore this claim and the government continues to collect taxes and provide goods and services to its citizens.

This volume presents an opportune occasion to examine why the search for a scientific method of evaluating government spending is bound to fail. It is fitting to begin with a number of prior questions concerning the evaluation

---

* I am indebted to Brain Wurts for research assistance in the preparation of this paper. I would like to thank Walter Hettich, Allan Maslove and Stan Winer for helpful comments on an earlier version of the paper. The opinions and conclusions are the sole responsibility of the author.

of government expenditures. Whose evaluation counts? What norms do we have against which to evaluate results? What quantitative indicators do we have that can be used to measure results? Is it meaningful to even try to "evaluate government spending"? The answers to these questions would seem to be especially crucial as part of the debate on government spending. After all, if we cannot agree on the norm of evaluation, then we have no theory of evaluation. If we cannot measure the indicators most appropriate for an agreed-upon norm, then we have no practical method of effecting an evaluation.

This paper is an attempt to provide a framework within which the above questions can be asked; it also provides some of the answers. In other words, I am asking the reader to join me, temporarily, in a quest for gold. I intend to point out the fool's gold along our path—the path of the analyst. In addition, I intend to suggest where the mother lode might be more likely found.

At the outset let me make clear that the focus of the paper is normative. The framework to be developed will encompass a number of "norms" or "goals" of government spending, and compare actual spending with these norms. Such a normative model tells us, "what ought to be"; it is a necessary foundation for any evaluation of government expenditures.

The framework of evaluation is developed in the following three sections of the paper. We first discuss the choice of the norm(s)—a choice which is intimately linked with some set of value judgments. It is the thesis of this paper that the existence of multiple norms in a complex society renders the attempt to derive a single-valued agreed-upon macro-evaluation of government spending futile. We then narrow our focus and discuss the choice of the indicators that could be used to approximate each separate norm or goal. Here too, the existence of complex sets of value judgments renders impossible the selection of one indicator only for each goal; rather, an array of indicators is necessary. Finally, we discuss the quantification of the indicators. This is primarily a technical task, to which the analyst can bring some expertise, but values held by the members of the community—including the analyst—intrude here as well.

Our conclusions are found in the fifth section. To anticipate the ending of our quest, the role of the analyst is limited to the modest, but not easy task of providing information on the possible indicators of the many norms held by members of the community. The analyst cannot evaluate government spending; he can provide input which will assist others in doing so. It is to the citizen and his handmaiden, the politician, that we must look for the truly creative act of evaluation of government spending.

SPENDING EVALUATION: CHOOSING THE NORMS

Private Market Expenditures

The genesis of any evaluative framework is rooted in some set of value judgments. This is as true for evaluating public spending as it is for evaluating a housewife's purchase of pepper squash. One can imagine a complex set of judgments being made as the housewife sets out to get "value for money." Let us suppose that, in terms of taste-bud preference, the wife, husband, son, and daughter in a family have, respectively, a mild liking, a mild dislike, strong dislike, and strong liking for pepper squash. The parents, in addition, place a high value on nutritional value for themselves and the children; and they place some value—the wife's outweighing the husband's—on the acquiring of a taste for pepper squash now in order to enhance their children's future health. Finally, the wife values the colour and texture combination of a meal (to feast the eye as well as the palate), and pepper squash scores high within the menu of Thursday's meal. Let us further suppose that, at Thursday's relative prices in the market place, the wife is observed buying the squash.

Given the inclination of most economists to prefer market outcomes over non-market outcomes, one might think that it would be relatively easy to evaluate the housewife's purchase. After all, if all markets are perfectly competitive, the outcome is an efficient one. However, beyond this tautological conclusion is a much more complex situation with respect to the evaluation of private expenditures. First, the evaluation scores depend very much on who is doing the judging: the daughter will rank the purchase very highly while the son will rank it very lowly—each brings a very different set of likes and dislikes to the task. Second, the evaluation scores depend on the consumption "norms" or "goals" that are taken into account by the evaluator. The children start off with one norm: taste. The parents agree on two other norms—health and future health—and the wife has a fourth norm—visual and textural attractiveness. The choice of any one of these norms is based on personal judgments, and if all members of the consuming unit do not agree on these norms, there is no way to reach a single-valued, agreed-upon evaluation.

If each member of the family hires a consultant to advise the member on an evaluation of the purchase, there will be four very different evaluations, one stemming from each set of value judgments underpinning the "norms" adopted by each member. It is even possible that the consultant hired by each parent might advise them that they ought not to take into account their children's health and future health; but since this is a a value judgment of the consultant, it would quickly be rejected by the clients.

It seems that the sole function of the consultant is to confirm his client's value judgments in writing. Is there a useful function the consultant can perform? Several come to mind: the consultant could analyse the nutritional value of pepper squash to determine if it does enhance health (and thus confirm the parent's belief); he could conduct tests or survey the relevant literature to determine if a taste for pepper squash can be acquired (thus confirming the wife's belief and strengthening the value her husband places on this strategy as a vehicle to enhance the future health of the children); he could comb the world to find an alternative vegetable that would score as high on health, future health, and attractiveness grounds and score higher on taste grounds for three of the four members of this family; finally, the consultant could survey prices in the market place to assure the family of the cheapest set of prices, given transportation costs.

These technical studies could be of some value. However, if it were determined that pepper squash (1) has high nutritional value, (2) the taste for which can be acquired by one helping every three weeks, (3) has no superior alternative on all four norms, and (4) the relative prices prevailing on Thursday were the true relative prices, then, the consultant can do no more than withdraw back into the shadows, leaving the four family members with four evaluations of private expenditures on pepper squash.

The above example has been drawn out in some detail because its relevance is crucial to the thesis of this paper. In the market for private expenditures it may be possible to quantify the efficiency with which various "norms" or "goals" are purchased. In those consuming units where purchasing power over some items is delegated to one or more members of the consuming unit, and where the members hold very different value judgments about the possible "norms," it is not possible to derive a single global evaluation that is satisfactory to all members of the unit. In fact, it is futile to even attempt such a task.

Public Market Expenditures

The above conclusions hold in the public market of government expenditures. It may be possible to quantify the efficiency with which various "norms" are purchased. It is not possible to derive a single measure of evaluation in a multiple norm community that is satisfactory to all members of the community.[1] In this sense, at least, the search for an ideal method of evaluation of government expenditures is a search for fool's gold.

The two problems inextricably linked with value judgments of the members of a spending unit that delegates purchasing authority permeate all aspects of public spending. First, the evaluation scores depend on who is

---

[1] It might be argued that no analyst in his right mind would attempt to (or has attempted to) derive such a measure. Even if the point is accepted, it still remains true that (1) many analysts while in the process of evaluating *one* dimension of government spending imply that the evaluation is more comprehensive and/or definitive; and (2) members of the community who place a high normative value on that selfsame dimension interpret the results as a comprehensive evaluation of government spending.

doing the judging. Not only are there many citizens who value differently the provision of any public good (in some models only the median voter is completely satisfied with the level of public spending—Buchanan 1970, p. 117-25), but there are two major sets of actors who derive personal benefits from the provision of public goods—politicians and bureaucrats. The politician can be expected to rank highly those public projects which generate relatively high benefit to *his* constituents at relatively low tax costs to *his* constituents (Downs 1957), independent of the aggregate benefits and costs of the project. Thus, the evaluation of public spending by politicians is most likely to be linked to the vote-getting attributes of the project. The bureaucrat, within the constraints imposed by the political process, can be expected to rank highly those public projects that both satisfy voter-generated demand and enhance his real wealth position—either through expanding the size of the bureau firm (Niskanen 1971), or by transforming the difference between revenues and costs into a non-monetary benefit.

As a result of these many different actors, each one of which could have a very different set of judgments with respect to a particular public project, it is not possible to devise a single evaluation standard that satisfies everyone. The value judgments each actor carries into the game determine his evaluation of the outcome of the game.

Second, the evaluation depends on the "norms" or "goals" which the many members of the community have. If one voter attempts to achieve an efficient level of public spending on a particular public project, whereas another voter seeks to alter the distribution of income through public spending on that same public project, then the two voters will have different frames of reference when it comes to evaluating the public spending on the project. These and other potential "norms" vary across voters and spring from differences in value judgments held by the different voters. Such differences in values or beliefs cannot be reconciled by any single-valued macro-evaluation technique.

There are many possible norms of government spending. To simplify matters I am going to concentrate on the three that cement together Musgrave's multiple budget theory of the public household (Musgrave 1959) and are most regularly referred to in the literature—allocative efficiency, distributive justice, and stabilization of income. Just these three norms render the task of evaluation a difficult one. Let us suppose for example, that we have a community with three voters who interest us. The Gladstonian voter places value solely upon allocative efficiency: if the only way to purchase certain goods—collective consumptions goods—is through a collective organization, called government, he wants to purchase his benefits at as low a tax cost as possible. The Keynesian voter values efficiency in public spending, but he also places a high value upon having a stable flow of income through time, uninterrupted by periods of inflation and massive unemployment of resources. This voter is prepared to direct taxing, spending, and monetary policy to purchase a stable flow of income.

The 'Ksanian voter puts some value on efficiency in public spending and stability in the flow of income over time, but he places a much higher value on an equitable distribution of income.[2] Such a voter does not necessarily desire an equal income for all, but he does desire—and is prepared to contribute resources to assist in achieving—a more equal sharing of the resources of the community. This voter is prepared to direct the taxing and spending policy of his government to purchase a more equal distribution of income.

In this simple example, the Gladstonian voter, the Keynesian voter, and the 'Ksanian voter each has a set of value judgments which form the basis of the "goals" or "norms" against which government spending and taxing is going to be judged. The norms, in other words, follow from the judgments and values of individual voters. No consultant can choose them; he can only hope that his analysis includes those norms which voters hold. For the thesis of this paper, it is sufficient if more than one "norm" is held; I am using three for illustrative purposes; the argument would in no substantive way be affected if individual voters held thirteen "norms." Assuming a consultant's analysis contains precisely those "norms" held by the voters, can a single-valued evaluation of government spending on a project be made that has unanimous voter support? The answer is no.

A given project with known benefits financed by a given tax with known incidence pattern cannot simultaneously meet the preferences of the Gladstonian voter, the Keynesian voter, and the 'Ksanian voter. If it met the norm of the Gladstonian voter it would score low on the evaluation scorecards of the Keynesian and the 'Ksanian voter. A consultant to the Gladstonian voter would evaluate the project highly; a consultant to the 'Ksanian voter would evaluate the project as a poor one. These consultant evaluations would, of course, do nothing more than articulate in writing the different value judgments of the voters.

In this simple example, the three voters would receive from the three consultants three different evaluations, each one of which would be correct for one voter. The consultants would not be able to reconcile their evaluations or derive a single, agreed-upon macro-evaluation because the source of the difference lies in the beliefs of the voters. In general, then, no single-valued evaluation can be derived to encompass the multiple goals of

[2] The analogy is drawn from the 'Ksan Indian tribes of British Columbia. The 'Ksan, along with many West coast Indian tribes, practised a form of "potlatch" wherein to achieve status and maintain the culture of the tribe a member gave away his belongings, gifts, and wealth to others during feasts that could last for as long as three months. The "potlatch," a massive giving of resources to others, is a precursor to a modern day redistribution of income. The "potlatch" which was the culmination of 'Ksan culture, was outlawed by the Canadian government in 1884.

allocative efficiency, distributive justice, and stabilization of income. It is diverse value judgments that prevent such an evaluation.

If there is to be any evaluation of public spending it has to be less comprehensive and more disaggregated—possibly geared to the separate norms held by the voters. It may be manageable to evaluate the attainment of the three goals separately; thus goal evaluation rather than government spending evaluation would become the object of the exercise.

GOAL EVALUATION: CHOOSING THE INDICATORS

With our sights narrowed from macro-evaluation to evaluation of individual goals it is possible to turn to a discussion of the indicators or measures of performance used to guide us in evaluating whether a particular goal is being achieved. Unfortunately, narrowing our evaluation task has not eliminated the intractable problem of value judgments. Values can—and often do—determine the choice of a particular indicator, regardless of how much we might like to believe that such a choice ought to be made on technical grounds alone.[3]

It is the mixture of technical judgments (Is the indicator conceptually sound? Are the data available? If the data are not available, is the indicator practical?) and the value judgment underlying the choice of indicator that strains the lines of communication and generates endless disputes between technicians and members of the community, and among them all. Members of the community continue to bring their complex normative judgments to bear on the "indicator" most appropriate to measure each separate goal. Technicians—economists, analysts, policy advisors and consultants—are members of the community as well, and most probably find it difficult to ignore their own complex value judgments and concentrate on technical questions alone.

Let us suppose, for example, that an economist has been asked to develop an indicator to be used as a measure of the successful (or unsuccessful) achievement of the distributive justice goal. He ignores the allocative efficiency and stabilization goals and focuses upon poverty as being the crux of the distributive justice goal. His study concludes by recommending that the distributive indicator be the number of families below the income level that permits a family to live "adequately" ($7100 for a family of four living in Ottawa in 1974).

Considerable criticism ensues—some of it couched in terms of technical requirements—and by the time it is over at least three other indicators have

---

[3] Note that this belief is a value judgment as well. The "we" I refer to here are economists, analysts, policy advisors, and consultants—all of whom can claim some competence of technical matters. "We" may believe in choosing indicators on technical grounds, not because the result is likely to be more objective or more conceptually sound, but because it is in our own interests to stress the service we sell. See, for example the discussion in Maslove (1975) pp. 481-84.

emerged as serious contenders. Those 'Ksanian voters who view poverty as the focus of the distributive justice goal but define it differently (for example, they believe poverty levels should permit a family to live in "decency and dignity" (Canada 1968), not just "adequately") recommend a poverty level of $8900 for the same family of four. Those 'Ksanian voters who believe distributive justice is defined in terms of the shares of income of all families throughout the distribution of income recommend that the distributive indicator be the decile distribution of income. In addition, the Gladstonian voters, who do not want to pay for any distributive justice at all, do have a view on the appropriate indicator to use: that indicator that minimizes any payment by them. Consequently, the Gladstonian voters define poverty as the focus of the distributive justice issue but argue that only the barest subsistence level ought to be considered (the poor ought to have minimal food, shelter, and clothing, but no recreation, travel, or entertainment): they recommend a poverty level of $3000.

Thus, an *apparently* technical task results in four contending indicators of distributive justice. And the source of contention is not technical differences of judgment among professionals but complex value judgment differences among members of the community. The same value judgments that render an agreed-upon single-valued evaluation of government spending impossible, also render a single indicator for each goal impossible. There will be, more likely, several indicators for each goal. Choice among such indicators cannot be made on technical grounds alone; rather it must be made on the basis of the underlying complex set of value judgments held by members of the community. A technician can present the results of using several indicators to capture each of the goals; he can rarely make a choice among the alternatives on strictly technical gounds. Members of the community and their elected politicians must make the ultimate choice.

The technician is restricted to presenting a set of indicators for each goal. It may be useful to suggest one such set of indicators for the three goals mentioned earlier. The specific array of indicators summarized in Table 1 is strictly for illustrative purposes. The value of the table lies in drawing together several indicators, each one of which has been used upon occasion to evaluate a spending or taxing decision. The argument of this paper is that since no one indicator captures the complex sets of value judgments held by members of the community, no one indicator can (1) be used to evaluate accurately government spending and taxing decisions, or (2) gain widespread support among members of the community, regardless of how popular it may be among technicians.

The array of indicators hints at the intractability of evaluating public expenditures. Twenty-four indicators present even the most ambitious of technicians with a formidable task, and for the most part each indicator also encompasses a set of value judgments.

Table 1
Evaluation Indicators

Goals	Allocation	Distribution: Income class.	Distribution: Regional	Stabilization
Indicators	1. Benefit-cost ratio 2. Net present value 3. Individual fiscal incidence ratios	4. Number below: a. subsistence poverty line (minimum needs) b. adequate poverty line c. decent poverty line 5. Income share comparisons of a. quintiles b. deciles 6. Gini coefficient 7. Group fiscal incidence ratios a. comprehensive b. partial	8. Regional per capita factor income 9. Regional per capita total income 10. Number below poverty line(s) across regions 11. Income share comparisons across regions 12. a. Regional fiscal incidence ratios b. Regional input-impact-related net benefit measure	12. Variations in aggregate indicators: unemployment and inflation 14. Fiscal policy:* a. qualitative b. fiscal leverage c. full employment surplus d. matching budgetary results with budgetary plans 3. econometric fore-casting model plus qualitative f. sources of funds for changing financial requirements

* We exclude monetary policy indicators because of the paper's exclusive focus on government spending and taxing

Since the technicians cannot narrow the array of indicators, it remains only to attempt to quantify them, thereby providing the data input which can form the basis of evaluation by politicians and other members of the community.

QUANTIFYING THE INDICATORS

With the quantification of the indicators the technician finally has some task to which his expertise can be brought. He can recommend and devise a conceptually sound methodological approach for quantifying the indicators of Table 1. This quantification is not without its own unique set of problems.

In the first place an indicator summarizes information. Of necessity, some information will be left out. Different indicators relating to the same goal or variable leave out different information. The excluded information may be more important to one group of voters than a different group of voters, and it is not possible for the analyses to discount such excluded information on technical grounds alone. Consequently, analysts should refrain from advocating a unique measure or indicator and be prepared to devise several indicators. This problem is not unique to the measure of evaluating goals of government spending, but it may be more pervasive than elsewhere.

Secondly, there are conceptual problems in connection with almost every indicator. In some instances these conceptual problems have become the basis for controversies in the literature revolving around the "theoretically correct" way of measuring the indicators in question. Benefit-cost analysis and net present value analysis (indicators 1 and 2) have given rise to professional differences of judgment over the proper treatment of unemployed resources, the correct discount rate, and whether or not to integrate distributional aspects into the analysis (Baumol 1969; Buchanan 1969; Eckstein 1961; Feldstein 1964 and 1964a; Harberger 1971; Haveman 1965 and 1969; Hettich 1971 and 1976; Marglin 1963 and 1965; Margolis 1969; Mishan 1971; Musgrave 1969; Prest and Turvey 1965; and Weisbrod 1968 and 1969). The latter controversy is especially significant since it points up the re-emerging problem of trying to make one indicator do two jobs—allocation and distribution—where the two jobs reflect diverse value judgments and yet may not be separable (see especially Hettich 1976 and Weisbrod 1969). Part of the conceptual problem is inextricably linked to complex value judgments which we have tried to abstract from in this section.[4] In short, the technician *has* to deal with conflicting value judgments in the choice and construction of social indicators.

---

[4] As Bromley reminds us we *cannot* ignore value judgments; specifically we cannot ignore the evaluation stance of economists in the debate whether or not to include distributive considerations in orthodox benefit-cost analysis (Bromley 1976). And while these judgments may add bite to a technician's argument, they do not aid in the purely technical objective task of measuring the indicator.

A disinterested observer might feel more sanguine over the attempt to exclude distributional issues from benefit-cost analyses if it were not for the fact that "exclude" seems to mean "ignore." The belief among many analysts that orthodox benefit-cost analysis provides *the* evaluation of public projects (thereby stressing efficiency to the exclusion of all other norms) plus the assertion that lump-sum income transfers can correct any resulting undesired distributive effects undermine the exaggerated claims that are made for the technique (Harberger 1971). As a result, some analysts have attempted to integrate distributional issues into the benefit-cost technique; and while considerable ingenuity has been demonstrated in devising distributive weights, the "expanded" analysis is still very much in its infancy.[5]

The attempt to resolve the discount rate controversy—whether a social rate of time preference or a private market rate ought to be used to discount future benefits and costs—has shown similar ingenuity. In the absence of a definite resolution of the issue, however, practitioners have used a range of interest rates to test the sensitivity of the results.[6]

Fiscal incidence analysis (indicators 3 and 7) has led to differences of professional judgment over the extent of the general equilibrium model required for the shifting hypotheses for several taxes (sales taxes, corporation income tax and, more recently, property tax), and the appropriate counterfactual (income base) against which to compare the distributive effect of the government sector (Aaron and McGuire 1970; Bird and Slack 1976; Bishop 1961; Meerman 1974; Peacock 1974; Peacock and Shannon 1968; and Reynolds and Smolensky 1977, chap. 2). The income base controversy is especially illuminating. In addition to encompassing strictly technical alternatives, it embraces a "conservative" alternative income concept (wherein the emergence of a government sector alters the original distribution of income in favour of lower-income families—Friedman 1962), and a "radical" alternative income concept (wherein the emergence of a government sector alters the original distribution of income in favour of higher-income families—Gordon 1972; Michelson 1970, Peppard 1976; and Sawers and Wachtel 1975). In other words, controversy over a conceptual issue masks very different models of the way in which society operates. These different "perceptions of how an economy behaves" should be made clear by the technician when presenting a particular indicator.

Regional income and distribution analysis (indicators 8 and 9) has led to conceptual problems concerning the appropriate delineation of an economic region (Atcheson, Cameron and Vardy 1974; Brewis 1969, chap. 3;

---

[5] See, for example, the excellent discussion of this issue in Hettich (1976).

[6] The Treasury Board in its latest advice to departments on the proper conduct of benefit-cost analysis (Treasury Board 1976, p. 26) recommends a range of 5 per cent (Helliwell *et al*. 1973) to 15 per cent around a median of 10 per cent (Jenkins 1973).

Cameron, Emerson and Lithwick 1974; and DREE 1969 and 1976), and the extent of potential conflict between regional per capita indicators and regional distribution indicators (Gillespie and Kerr 1977; Maslove 1975; and Usher 1975). Policies to equalize regional per capita incomes may result in a more unequal national distribution of income or more unequal distribution of income within certain regions. In such a case the difficulty is whether per capita income is a conceptually sound indicator of a regional distribution objective.

The above are but a few of the conceptual problems of several of the indicators of Table 1. They should be sufficient to caution the reader against a too hasty acceptance of the quantification of the indicators as a straightforward technical task: value judgments permeate the analysis, even at the micro-level of a social indicator.

In addition to these conceptual problems there are technical and data availability problems that render the task of quantification even more problematical. Benefit-cost and net present value analyses require estimates of demand curves to calculate consumer surplus values. However, demand curves for collective-consumption goods cannot be directly measured precisely because of their collective consumption properties and the non-revelation of preferences for such goods (Musgrave 1969). In some instances shadow prices can be estimated (Margolis 1969); in others, considerable ingenuity has gone into devising estimates of minimum benefits (Fisher, Krutilla and Cicchetti 1972); but in many cases the analyst must forgo hard data on benefits. Without solid benefit data benefit-cost analysis loses its intellectual appeal.[7]

Poverty line analysis of income distribution (indicator 4) requires a count of those persons with resources below the specified poverty income levels. After "subsistence," "adequate," and/or "decent" poverty income levels have been *defined*, the technician must translate these definitions into many different family situations (urban, rural; number of persons in the family; possibly age of head) and then count the poor.

Considerable progress has been made in an accurate count of the poor since the Economic Council first drew attention to the extent of poverty in Canada (Economic Council of Canada 1968, chap. 5). Statistics Canada's annual surveys of incomes of families include a count of low-income families, by several cross-classifications (Podoluk 1968; and Statistics Canada 1976). The low-income cut-off point approximates an "adequate" poverty line (indicator 4b).

---

[7] This is not to argue that benefit-cost analysis has no contribution to make. It can "prove an *appropriate framework for thinking* about the problems and it may provide some useful information" (Economic Council of Canada 1971, p. 53, emphasis added). In addition it "can, of course, be useful in helping legislators ask proper questions" (Bromley 1976, p. 832). These are modest contributions by anyone's standards.

However, it is still correct that a technician is virtually limited to these data, thus eliminating a possible count of those below a "subsistence" poverty level (indicator 4a) and those below a "decent" poverty level (indicator 4c). With data availability constraining the feasible poverty line indicators to one, it should not be surprising that many consumer-voters view the publication of official poverty statistics with sceptical disbelief. Those who *believe* poverty ought to be defined as a bare subsistence level will reject the indicator as an overcount of the poor (Ontario Economic Council 1976, pp. 5-7). Those who *believe* poverty ought to be defined as just below a more "decent and dignified" standard of living or in relative terms will reject the indicator as an underestimate of the number of poor (Croll 1971 and Adams *et al.* 1971).In this instance, the lack of data consistent with alternative views of poverty may undermine support for publishing or making available the existing series rather than stimulating support for making available a number of alternative series.

Fiscal incidence methodology has numerous technical difficulties and data problems as well as the conceptual controversies noted earlier. The data for most counter-factual income concepts are not available, nor, as it may turn out, can they be obtained. For example, the "radical" counter-factual requires the factor income distribution prior to the government's effect in altering the distribution of factor income in favour of the higher-income families. Such an income distribution does not exist, and it is not possible, given the present state of the art of fiscal incidence studies, to derive a rough estimate. A similar impasse exists in using the "conservative" counter-factual. This may account for the use of the orthodox "neutral" counter-factual, not only in studies which are premised upon an original "neutral" government (Gillespie 1965, 1966, 1976 and 1977; Johnson 1968; and Dodge 1975) but also in studies that are strongly critical of the "neutral" government approach (Reynolds and Smolensky 1977; and Peppard 1976).

It should not be surprising, therefore, that many consumer-voters are sceptical of the results of the fiscal incidence indicator. Those who have a "radical" view will judge that the orthodox methodology *overstates* the degree of redistribution towards lower-income families (Michelson 1970). Those who have a "conservative" view will judge that the methodology *understates* the degree of redistribution towards lower-income families. The lack of available data, therefore, may not only limit the methodology but may also result in a rejection of the results for invalid reasons (i.e., the methodology does not conform with a different view of the economy).

The stabilization indicators are also fraught with technical problems. The aggregate indicators (13) have to be defined for policy purposes. For example, the distinction between voluntary and involuntary unemployment, hardly ever clear in a practical measurement sense and rendered even more opaque in "natural" rate of unemployment models, must be translated into clear guide-posts for the counting of the unemployed. The alternative of the technician is to adjust upward—sometimes on the basis of an explicit macro-model, from which such predictions can be extracted—the "frictional" and/or "natural" rate of unemployment.[8] Such an alternative is open to technical criticism depending on the model chosen and the reliability of the results. It is also open to critical scepticism by members of the community—especially those who are opposition members of Parliament—who may suspect that the fact of rising aggregate unemployment is being *defined away* by such a technical adjustment. For possibly the very same reasons the alternative is less critically accepted by those consumers who suspect that involuntary unemployment has been defined *too generously* in the past, and by a government which considers that it has to tackle unemployment in some way. Again, the technical merits of the measure of involuntary unemployment and of changes in the method of measurement become linked to debate which is heavily based on different complex sets of value judgments and/or views of the operation of the economy held by different members of the community.

The technical problems are equally formidable in the case of fiscal performance indicators (14a through 14f). The full employment surplus indicator of fiscal performance for stabilization purposes (indicator 14c) is built upon an estimate of what government revenues and government expenditures would be at a full employment level of output. Therefore, for practical reasons it is necessary to be specific about the assumed rate of voluntary unemployment (equals full employment) in the economy. The Royal Commission on Taxation took 3.5 per cent voluntary unemployment for its study of the 1950s and early 1960s; the Economic Council of Canada took 3.8 per cent for its study of the late 1960s and early 1970s. A technician could be critical of both measures, especially if he placed confidence in recent estimates of the natural rate of unemployment of 5.1 per cent.[9]

The technician has to use tax multipliers and expenditure multipliers to estimate budgetary revenues and expenditures at full employment. The former exist—although there may be considerable variance in the value of a particular multiplier due to different estimating models. However, such multipliers rarely, if ever, allow for the possibility that as the system moves

---

[8] For example, a number of recent attempts have been made to estimate the effect of the unemployment insurance revisions of 1971 on the rate of unemployment. The findings that the revisions would appear to have added from 0.6 per cent (Green and Cousineau 1977) to 0.8 per cent (Grubel, Maki and Sax 1975) to the unemployment rate imply a "natural" rate of unemployment of approximately 5.1 per cent to 5.3 per cent.

[9] The Brookings Institution recently provided an alternative potential GNP estimate for *three* different estimates of unemployment at full employment, the data that are crucial for calculations of full employment surplus estimates. One series is based on a 4 per cent unemployment rate at potential full employment through 1955-1981; one series is based on an unemployment rate that varies from 4 per cent in 1955 to 4.9 per cent in 1976-1980 and 4.8 per cent in 1981; and a third series is based on an unemployment rate that varies from 4 per cent in 1955 to 4.9 per cent in 1977-1978 and 5.0 per cent in 1980-1981 (Pechman 1977, pp. 422-23).

closer to full employment, legislated changes in the tax structure—tax loopholes, now more respectfully defined as tax expenditures—could alter the multiplier value even more. This problem is not solely a technical difficulty since its resolution would require some model of government behaviour to be built into the analysis. On the expenditure side of the budget, the standard assumption is that the level of government expenditures at full employment would be identical to existing spending—with the exception of unemployment insurance benefits. It is recognized that this is an over-simplification, but little attempt is made to correct it. The reasons are twofold: expenditure multipliers would require a complete model of government behaviour, a model which we do not have; and empirical estimates of expenditure multipliers do not exist. Thus, the lack of information and data inhibits the technician constructing full employment surplus estimates (Gramlich, 1968; Okun and Teeters 1970; and Smyth 1970).

In short, conceptual, technical, and data availability problems render the task of quantification of indicators difficult. The pervasive role of complex sets of value judgments cannot be ignored, even in this apparently "technical" and limited task.

CONCLUSIONS

In concluding let me draw together the main observations of this paper. First, the existence of multiple norms in a complex dynamic society renders impossible the attempts to derive a single-valued agreed-upon macro-evaluation of government spending. Second, even when we narrow our evaluation perspective to the focus of separate norms, the same set of complex value judgments renders impossible the selection of one optimal or ideal indicator for each goal. There will, more likely, be several indicators for each goal and any choice among them can rarely be made on technical grounds alone. Rather, the choice of any indicator has to be made on the basis of the underlying set of value judgments.

Third, the attempt to quantify the indicators is fraught with frustration. There are major unresolved conceptual issues at the core of virtually every indicator. There are technical and data availability aspects which are sufficiently serious to raise serious doubts in several cases as to whether a conceptually sound indicator can be derived.

Finally, the very nature of some of the conceptual, technical, and data problems is such as to perpetuate the tension among technicians and members of the community. And rightly so, because these problems guide, cajole, and back the technician into choices which conflict with some (while coinciding with other) complex sets of beliefs and value judgments.

I do not conclude from this that the technician should pack his tool-kit and go to the beach.[10] There is much to be done—some of which I have endeavoured to hint at above—before we technicians join the unemployed. In the appendix I examine the extent to which the evidence on indicators pertinent to an evaluation of federal spending in Canada has progressed.

Rather, I conclude that we technicians must eschew the search for the global evaluation indicator and demonstrate considerably more humility in presenting the results of goal evaluation with a particular indicator (and the twenty-four indicators of Table 1 are not all-encompassing). A technician's estimate of a particular indicator may be based on a sound theoretical model, devoid of technical difficulties, and still receive short shrift by a public which places a higher value on other goals and indicators. Since there is no one indicator of public spending evaluation, the public will be required to work through the many indicators of diverse aspects of public spending, apply its set of value judgment trade-offs, and send signals through its elected politicians for change where change is required.

The technician's role in the whole evaluation of public spending debate is a limited one. He can provide the evidence for the indicators—the array of indicators, as expended, of Table 1. He can make explicit the value judgments underlying the particular social indicator. He can explain the measures clearly to politicians and other members of the community.

However, the creative work in the evaluation of public spending has to be left to others. It is to each member of the community with his or her personal evaluation of the norms of public spending, and to the politician who competes with other politicians for the privilege of reconciling the conflicting personal evaluation of the many voters, that we must look for the truly creative act of evaluation.

---

[10] This is not the place to enter into a detailed discussion of the role of the economic advisor in decision making within the government sector. Suffice it to say that there is considerable difference of opinion within the discipline on this matter—almost all of which differs from the position developed here (see, for example, the quotations at the beginning of the paper: Bromley 1976, p. 812; Paquet 1971, p. 55; and Peacock 1977, pp. 22-23).

Appendix

There exists a modest published output for most of the indicators of Table 1. The relevant studies are listed in appendix Table A-1. A brief commentary may be in order.

The paucity of quantitative evidence for indicators 1 and 2, the indicators apparently preferred by many technicians for evaluating government spending, may be a result of restricting the table to published (or widely available) research. Benefit-cost analysis as an aid in choosing among competing projects is primarily an *ex ante* instrument. As such there may exist, deep in the bowels of the federal bureaucracy, numerous examples of benefit-cost research that have preceded the decision to embark on a new spending project (possibly, Matador 1977). We have been able to detect only the *ex post* studies, done by non-bureau technicians for after-the-fact evaluations.

It is interesting to note that when Treasury Board updated its benefit-cost guide in 1976 (from Sewell *et al.* 1962 to Treasury Board 1976) it did *not* include among its three examples of the technique any example drawn from a federal program or project. Could it be that there do not exist within the federal bureaucracy *ex ante* benefit-cost analyses that come close to conforming with the Treasury Board's view of an appropriate benefit-cost analysis? Or could it be that departments were not prepared to allow publication of their benefit-cost analysis for fear of adverse critical appraisal? Or could it be that, contrary to popular belief, benefit-cost analysis is not carried out within the federal bureaucracy? In any case, the Gladstonian voter, lacking the information, has little substantive basis for joy or depression over federal spending and taxing decisions.

The beginning of an extensive field of quantitative evidence on the distribution of income has done little to minimize confusion over the meaning of the indicators—especially the confusion between indicators 4 and 5 (or 6). When the distribution problem is defined as a problem of poverty in some absolute sense (indicators 4a and 4b and possibly 4c), then the number of poor will decline as real incomes rise over time. This is precisely what occurs: the proportion of low-income families has declined through time (Statistics Canada 1976, p. 100). When the distribution problem is defined as a problem of extreme differences in relative incomes, or when poverty is defined in some relative sense (indicators 5 and 6 and possibly 4c), then the number of poor may decline or increase as real incomes increase through time, and the share of income of the poorest group may rise, fall, or remain constant. In fact, the poorest 20 per cent of Canadian family units has experienced virtually no change in their share of total income since 1951 (Horner and MacLeod 1975; and Love and Wolfson 1976).

Table A-1
Evaluation Indicators: Evidence for Canadian Federal Government Spending and Taxing Decisions*

Goals	Allocation	Distribution: Income Class.	Distribution: Regional	Stabilization
Evidence on Indicators	1. Diamond Sewell (1971)	4a. Ontario Economic Council (1976: 739-43)	8. DREE (1976); Economic Council of Canada (1977: chap. 4)	13. —
	2. Borins (1978); National Energy Board (1977); Matador (1977); and Transport Canada (1978)	4b. Statistics Canada (1976); Economic Council of Canada (1968: chap. 5); Croll (1971); Lithwick (1971); and Canadian Council on Social Development (1975)	9. Brewis (1969); Chernick (1966); Economic Council of Canada (1965, 1975, 1977: chap. 4) and McInnes (1968)	14a. Gordon (1966); and Will (1967)
	3. —	4c. Adams *et al*. (1971)	10,11 —	14b. Auld (1969)
		5,6 Economic Council of Canada (1976); Health and Welfare Canada (1977); Henderson and Rowley (1977); Horner and MacLeod (1975); Love and Wolfson (1976); Statistics Canada (1976)	12a. Gillespie (1977) and Gillespie and Kerr (1977)	14c. Economic Council of Canada (1970, 1973, 1974); Royal Commission on Taxation (1966)
		7a. Dodge (1975); Gillespie (1966, 1976, 1977); and Johnson (1968)	12b. Co-ordination Group (1977); Government of Canada (1977); Gouvernement du Québec (1977, 1977a); Government of Ontario (1977); Statistics Canada (1977)	14d. Gillespie (1973, 1977a)
		b. Cloutier (1978); Deutsch (1968); Gillespie (1978); Manga (1978); and Maslove (1972)		14e. Curtis and Kitchen (1975); Jump and Wilson (1974, 1975, 1976)
				14f. Christofides (1977)

* The array of studies is organized according to the array of indicators in text Table 1; the table is meant to include all published studies, but some references may have been missed.

The facts underlying these two indicators are not different. What is different is the set of value judgments that are captured by indicator 4b and the set that are captured by indicator 5. Disagreement over the indicators reflects disagreement over the underlying value judgment sets, although such disagreement often surfaces as complete confusion. A recent example is illuminating: Mr. Trudeau was asked a question about the distribution of income with the focus of the questioner on indicator 5; the prime minister responded with the focus on indicator 4b (Canada 1978, pp. 4683-84). Thus the questioner could (correctly) point to the effectiveness of federal government programs in improving the relative economic position of low-income families during the 1970s, and the prime minister could (correctly) point to the decline in the number of poor in absolute terms during his tenure in office. A disinterested observer witnessing this debate could be confused into believing the disagreement was over technical matters (the soundness of measure, the derivation of a number, etc.), rather than over fundamental differences in value judgments with respect to the distribution of income in Canada.

Indicator 7 expands upon the results of indicators 5 and 6 by allowing for the effect of government expenditures, transfers, and taxes on the distribution of income. The results over time, while still the object of some controversy, seem clear: the poorest fifth of families did *not* improve their economic position relative to the richest fifth of families. In fact, both the poorest and richest families gained relative to the median income families, with an average highest-income family gaining approximately three times as much as an average poorest-income family (Gillespie 1977).

In short, the 'Ksanian voter has no basis for joy over federal spending and taxing decisions during the 1960s and 1970s (except for the 'Ksanian voter who embraces indicator 4b).

Quantitative evidence on indicators related to the regional impact of federal spending and taxing decisions is modest. This is especially surprising given the importance of regional issues within the Canadian confederation and given the importance which numerous federal governments have stated that regional issues merit. We even have a federal department in charge of 'regional economic expansion'.

Recent political developments in Quebec virtually guarantee that this research field will not remain unploughed for long. Part of the debate between Ottawa and Quebec over, 'Does Quebec benefit from confederation?' centres around the provincial distribution of federal taxes and government expenditures. The first crop of studies directed to answering this question have used, almost exclusively, indicator 12b which measures the immediate impact or location of the government spending and the tax collection (Co-ordination Group 1977; Government of Canada 1977; Gouvernement du Québec 1977, 1977a; Government of Ontario 1977). This input-impact-related method leads to the type of problem illustrated by the treatment of customs duties in the Quebec study (Gouvernement du Québec 1977, 1977a). With the methodology measuring tax collections rather than tax incidence (who bears the burden of the tax), the bulk of Canadian customs duties which are collected at the port of Montreal were *all* assigned to Quebec. Thus it appeared that families in Quebec bore a large tax burden. However, all consumers of import-competing commodities bear the incidence of customs duties, and when the methodology allows for this (as indicator 12a does), the burden of the tax is distributed among *all* families in Canada in relation to their consumption expenditures: families in Quebec bear a smaller share of the burden of this tax.

There are numerous differences between indicator 12a and indicator 12b. The major methodological difference captures a very different way of how an economy behaves. Indicator 12b focuses upon the factor input side of the location of government spending or taxing; if a bureaucrat's income rises because he is producing more social services, the benefit is attributed to the bureaucrat, who most likely lives in Ottawa. Indicator 12a focuses upon the output side of the benefit of government spending, and the benefit of the above spending would go to the consumer of social services, who most likely lives outside of Ottawa. These differences in methodology between factor receivers as beneficiaries and output consumers as beneficiaries account for the major differences in findings between indicator 12a studies and indicator 12b studies.

The evaluations of fiscal policy performance found in the 14a through 14e indicators agree on one important count: fiscal policy was not very effective in achieving stabilization of income during the post-war period. From Gordon's pessimistic evaluation (adequate 20 per cent, perverse 40 per cent of the time—Gordon 1966) through the Curtis and Kitchen optimistic appraisal (adequate 45 per cent of the time—Curtis and Kitchen 1975), fiscal policy emerges as a tool that, in the hands of the federal government, did not live up to expectations.

In short, the Keynesian voter has no basis for joy over federal spending and taxing decisions of the post-war period.

In concluding this appendix, I note that three value judgments have crept into the analysis. Let me rephrase these observations as my judgment as to how voters might be expected to react. I believe that the Gladstonian voter, lacking any information on efficiency, would have no basis for exultation or resentment over federal spending and taxing decisions. I believe that the 'Ksanian voter and Keynesian voter would have no basis for joy and some for disappointment over these spending and taxing decisions of the federal government.

Given my restricted framework, there may, of course, be other voters who are ebulliently joyful or deeply resentful over federal spending and taxing. I have analysed three norms only and examined the possible indicators associated with these norms. However, there may be other norms for which some voters are ebulliently joyful (after all the Liberal government has been re-elected several times!). I may have, in the analysis of Tables 1 and A-1, as I cautioned the reader earlier, inadvertently stumbled upon another cache of fool's gold, rather than the real thing.

Chapter Three

A Budget for All Seasons? Why the Traditional Budget Lasts

by
*Aaron Wildavsky**

INTRODUCTION

Almost from the time the caterpillar of budgetary evolution became the butterfly of budgetary reform, the line-item budget has been condemned as a reactionary throw-back to its primitive larva. Budgeting, its critics claim, has been metamorphized in reverse, an example of retrogression instead of progress. Over the last century, the traditional annual cash budget has been condemned as mindless, because its lines do not match programs; irrational, because they deal with inputs instead of outputs; short-sighted, because they cover one year instead of many; fragmented, because as a rule only changes are reviewed; conservative, because these changes tend to be small; and worse. Yet despite these faults, real and alleged, the traditional budget reigns supreme virtually everywhere, in practice if not in theory. Why?

The usual answer, if it can be dignified as such, is bureaucratic inertia. The forces of conservatism within government resist change. Presumably the same explanation fits all cases past and present. How, then, can we explain why countries like Britain departed from tradition in recent years only to return to it? It is hard to credit institutional inertia in virtually all countries for a century. Has nothing happened over time to entrench the line-item budget?

The line-item budget is a product of history, not of logic. It was not so much created as evolved. Its procedures and its purposes represent accretions over time rather than propositions postulated at a moment in time. Hence, we should not expect to find them either consistent or complementary.

Control over public money and accountability to public authority were among the earliest purposes of budgeting. Predictability and planning—knowing what there will be to spend over time—were not far behind. From

* This paper grew out of my collaboration with Hugh Heclo on the second edition of *The Private Government of Public Money*. I wish to thank him, James Douglas, Robert Hartman, and Carolyn Webber for critical comments.

the beginning, relating expenditure to revenue was of prime importance. In our day we have added macro-economic management, intended to moderate inflation and unemployment. Spending is varied to suit the economy. In time the need for money came to be used as a lever to enhance the efficiency or effectiveness of policies. He who pays the piper hopes to call the tune. Here we have it: budgeting is supposed to contribute to continuity (for planning), to change (for policy evaluation), to flexibility (for the economy), and to rigidity (for limiting spending).

These different and (to some extent) opposed purposes contain a clue to the perennial dissatisfaction with budgeting. Obviously no process can simultaneously provide continuity and change, rigidity and flexibility. And no one should be surprised that those who concentrate on one purpose or the other should find budgeting unsatisfactory; or that, as purposes change, these criticisms should become constant. The real surprise is that traditional budgeting has not been replaced by any of its outstanding competitors in this century.

TRADITIONAL BUDGETING vs ALTERNATIVES

If traditional budgeting is so bad, why are there not better alternatives? Appropriate answers are unobtainable, I believe, so long as we proceed on this high level of aggregation. So far as I know, the traditional budget has never been compared systematically, characteristic for characteristic, with the leading alternatives.[1] By doing so we can see better which characteristics of budgetary processes suit different purposes under a variety of conditions.

Why, again, if traditional budgeting does have defects, which I do not doubt, has it not been replaced? Perhaps the complaints are the clue: just what is it that is inferior for most purposes and yet superior over all?

The ability of a process to score high on one criterion may increase the likelihood of its scoring low on another. Planning requires predictability and economic management requires reversibility. Thus, there may well be no ideal mode of budgeting. If so, this is the question: do we choose a budgetary process which does splendidly on one criterion but terribly on others? Or, do we opt for a process that satisfies all these demands even though it does not score brilliantly on any single one?

A public sector budget is supposed to ensure accountability. By associating government publicly with certain expenditures, opponents can ask questions or contribute criticisms. Here the clarity of the budget presentation—linking expenditures to activities and to responsible officials—is crucial. As a purpose, accountability is closely followed by control: are the authorized and appropriated funds being spent for the

---

[1] But, for a beginning, see Allen Schick (1966).

designated activities? Control (or its antonym "out of control") can be used in several senses: are the expenditures within the limits (a) stipulated or (b) desired? While a budget (or item) might be "out of control" to a critic who *desires* it to be different, in our nomenclature *control* is lacking *only* when limits are stipulated and exceeded.

Budgets may be mechanisms of efficiency—doing whatever is done at least cost, or getting the *most* out of a given level of expenditure—and/or of effectiveness—achieving certain results in public policy such as improving health of children or reducing crime.

In modern times, budgeting has also become an instrument of economic management and of planning. With the advent of Keynesian economics, efforts have been made to vary the rate of spending so as to increase employment in slack times or to reduce inflation when prices are deemed to be rising too quickly. Here (leaving out alternative tax policies), the ability to increase and decrease spending in the short run is of paramount importance. For budgeting to serve planning, however, predictability (not variability) is critical. The ability to maintain a course of behaviour over time is essential.

As everyone knows, budgeting is not only an economic but also a political instrument. Since inability to implement decisions nullifies them, the ability to mobilize support is as important as making the right choice. So, too, is the capacity to figure out what to do, that is, to make choices. Thus, the effect of budgeting on conflict and calculation—the capacity to make and support decisions—must also be considered.

Traditional budgeting is annual (repeated yearly), incremental (departing marginally from the year before). It is conducted on a cash basis (in current dollars). Its content comes in the form of line-items (such as personnel or maintenance). Alternatives to all these characteristics have been developed and tried, though never, as far as I know, with success. Why this should be so, despite the obvious and admitted defects of tradition, will emerge as we consider the criteria each type of budgetary process has to meet.

UNIT OF MEASUREMENT: CASH OR VOLUME

Budgeting can be done not only in terms of cash but also in terms of volume. Instead of promising to pay so much in the next year or years, the commitment can be made in terms of operations to be performed or services to be provided. Why might someone want to budget in terms of volume (or constant currency)? To aid planning: if public agencies know that they can count, not on variable currency, but on what the currency can actually buy, that is, on a volume of activity, they can plan ahead as far as the budget runs. Indeed, if one wishes to make decisions now which could be made at future periods, so as to help assure consistency over time, estimates based on stability in the unit of effort—so many applications processed or such a level of services provided—are the very way to go about it.

So long as purchasing power remains constant, budgeting in cash or by volume remains a distinction without a difference. But should the value of money fluctuate (and, in our time, this means inflation), the public budget must absorb additional amounts so as to provide the designated volume of activity. Budgeters lose control of money because they have to supply whatever is needed. Evidently, given large and unexpected changes in prices, the size of the budget in cash terms would fluctuate wildly. Evidently, also, no government could permit itself to be so far out of control. Hence, the very type of stable environment which budgeting by volume is designed to achieve turns out to be its major unarticulated premise. Given an irreducible amount of uncertainty in the system, not every element can be stabilized at one and the same time. Who, then, will enjoy stability and who will bear the costs of change?

The private sector and the central controller pay the price for budgeting by volume. Budgeting by volume is, first of all, an effort by elements of the public sector to invade the private sector. What budgeting by volume says, in effect, is that the public sector will be protected against inflation by getting its agreed level of services before other needs are met. The real resources necessary to make up the gap between projected and current prices must come from the private sector in the form of taxation or interest for borrowing. In other words: for the public sector volume budgeting is a form of indexing against inflation.

Within the government, spending by agencies will be kept whole. The central budget office bears the brunt of covering larger expenditures and takes the blame when the budget goes out of control, that is, rises faster and in different directions than predicted. In Britain, where budgeting by volume went under the name of the Public Expenditure Survey, the Treasury finally responded to years of severe inflation by imposing cash limits, otherwise known as the traditional cold-cash budget. Of course, departmental cash limits include an amount for price changes, but this is not necessarily what the Treasury expects so much as the amount it desires. The point is that the spending departments have to make up deficits caused by inflation. Instead of the Treasury handing over the money automatically, as in the volume budget, departments have to request it—and their requests may be denied. The local spenders, rather than the central controllers, have to pay the price of monetary instability.

Effects of Inflation

Inflation has become not only an evil to be avoided but a (perhaps *the*) major instrument of modern public policy. Taxes are hard to increase and benefits virtually impossible to decrease. But similar results may be obtained through inflation, which artificially elevates the tax brackets in which people find themselves and decreases their purchasing power. Wage increases which cannot be contested directly may be nullified indirectly (and the real burden of the national debt reduced), without changing the ostensible amount, all by means of inflation. The sensitivity of budgetary forms to inflation is a crucial consideration.

From all this, it follows that budgeting by volume is counter-productive in fighting inflation because it accommodates price increases rather than encouraging the struggle against them. Volume budgeting may maintain public sector employment at the expense of taking resources from the private sector, thus possibly reducing employment there. There can be no doubt, however, that volume budgeting basically serves counter-cyclical purposes because the whole point is that the amount and quality of service do not vary over time, rather than going up or down in accordance with short-run economic conditions.

How does volume budgeting rate as a source of policy information? It should enable departments to understand better what they are doing since they are presumably doing the same thing over the entire period of the budget. But volume budgeting does poorly as a method of instigating change. For one thing, the money is guaranteed against price changes, so there is less need to please outsiders. For another, volume budgeting necessarily leads to interest in internal affairs (how to do what one wishes), rather than to seeking external advice (whether there are better things one might be doing). British departments unwilling to let outsiders evaluate their activities are hardly going to be motivated by guarantees against price fluctuations.[2]

TIME SPAN: MONTHS, ONE YEAR, MANY YEARS

Multi-year budgeting has long been proposed as a reform to enhance rational choice by viewing resource allocation in a long-term perspective. Considering one year, it has been argued, leads to short-sightedness—only the next year's expenditures are reviewed; overspending—because huge disbursements in future years are hidden; conservatism—incremental changes do not open up larger future vistas; and parochialism—programs tend to be viewed in isolation rather than in comparison to their future costs, in relation to expected revenue. Extending the time span of budgeting to three or five years, it is argued, would enable long-range planning to overtake short-term reaction and substitute financial control for merely muddling through. And the practice of stepped-up spending to use up resources before the end of the budgetary year would decline in frequency.

Much depends, to be sure, on how long budgetary commitments last. The seemingly arcane question of whether budgeting should be done on a cash or on a volume basis will assume importance if a nation adopts multi-year budgeting. The longer the term of the budget, the more significant

---

[2] Hugh Heclo and Aaron Wildavsky (1979, forthcoming).

inflation becomes. To the extent that price changes are automatically absorbed into budgets, a certain volume of activity is guaranteed. To the extent that agencies have to absorb inflation, their real level of activity declines. Multi-year budgeting in cash terms diminishes the relative size of the public sector, leaving the private sector larger. Behind discussions of the span of the budget, the *real* debate is over the relative shares of the public and private sectors—which one will be asked to absorb inflation and which one will be allowed to expand into the other.

A similar issue of relative shares is created within government by proposals to budget in *some* sectors for several years, and, in others, for only one year. This poses the question of which sectors of policy are to be exposed to the vicissitudes of life in the short term, and which are to be protected from them. Like any other device, multi-year budgeting is not neutral but distributes indulgences differently among the affected interests.

Of course, multi-year budgeting has its positive aspects—if control of expenditure is desired, for instance. A multi-year budget makes it necessary to estimate expenditures further into the future. The old tactic of the camel's nose—beginning with small expenditures while hiding larger ones which will arise later on—is rendered more difficult. Still, as the British learned, ''hard in'' often implies an even harder out. Once an expenditure gets into a multi-year projection, it is likely to stay in because it has become part of an interrelated set of proposals which might be expensive to disrupt. Besides, part of the bargain struck when agencies are persuaded to estimate as accurately as they can is that they will gain stability, that is, not be subject to sudden reductions according to the needs of the moment. Thus, control in a single year may have to be sacrificed to maintain limits over the multi-year period. And, should there come a call for cuts to meet a particular problem, British experience shows that reductions in future years (which are always ''iffy'') are easily traded for maintenance of spending in the all-important present. Moreover, by making prices more prominent due to the longer time period involved, large sums may have to be supplied in order to meet commitments for a given volume of services in a volatile world.[3]

Suppose, however, that it were deemed desirable significantly to reduce some expenditures in order to increase others. Due to the built-in pressure of continuing commitments, what can be done in a single year is extremely limited. Making arrangements over a three- to five-year period (with constant prices, five per cent a year for five years, compounded, would bring about a one-third change in the budget) would permit larger changes in spending, to be effected in a more orderly way. This is true; other things, however—prices, priorities, politicians—seldom remain equal. While the British were working under a five-*year* budget projection, prices and production could hardly be predicted for five *months* at a time.

---

[3] *Ibid*.

As Robert Hartman put it, "there is no absolutely right way to devise a long-run budget strategy."[4] No one knows how the private economy will be doing nor what the consequences will be of a fairly wide range of targets for budget totals. There is no political or economic agreement on whether budget targets should be expressed in terms of levels required for full *employment*, for *price stability*, or for *budget balancing*. Nor is it self-evidently desirable either to estimate where the economy is going and to devise a governmental spending target to complement the estimate, or to decide what the economy *should be* doing and to budget in order to encourage that direction.

In any event, given economic volatility and theoretical poverty, the ability to outguess the future is extremely limited. Responsiveness to changing economic conditions, therefore (if that were the main purpose of budgeting), would be facilitated best by a budget calculated in months or weeks rather than years. Such budgets do exist in poor and uncertain countries. Naomi Caiden and I have called the process "repetitive budgeting," to signify that the budget may be made and re-made several times during the year.[5] Because finance ministries often do not know how much is actually in the nation's treasury or what they will have to spend, they delay making decisions until the last possible moment. The repetitive budget is not a reliable guide to proposed expenditure, but an invitation to agencies to "get it if they can." When economic or political conditions change (which is often), the budget is re-negotiated. *Adaptiveness* is indeed maximized but *predictability* is minimized. Conflict increases because the same decision is re-made several times each year. Agencies must be wary of each other because they do not known when next they will have to compete. Control declines, partly because frequent changes make the audit trail difficult to follow, and partly because departments seek to escape from control so as to re-establish a modicum of predictability. Hence they obfuscate their activities (thus reducing accountability), and actively seek funds of their own in the form of earmarked revenues (thus diminishing control). Both efficiency and effectiveness suffer. The former is either unnecessary (if separate funds exist) or impossible (without continuity), while the latter is obscured by the lack of relationship between what is in the budget and what happens in the world. Drastically shortening the time frame wreaks havoc with efficiency, effectiveness, conflict, and calculation. But if immediate responsiveness is desired, as in economic management, the shorter the span the better.

---

[4] Robert A. Hartman (1978), p. 312.

[5] Naomi Caiden and Aaron Wildavsky (1974).

CALCULATION: INCREMENTAL OR COMPREHENSIVE

Just as the annual budget on a cash basis is integral to the traditional process, so also is the budgetary base—the expectation that most expenditures will be continued. Normally, only increases or decreases to the existing base are considered in any one period. If such budgetary practices may be described as incremental, the main alternative to the traditional budget is one which emphasizes comprehensive calculation. The main modern forms of the latter are planning, programming and budgeting (PPB) and zero-base budgeting (ZBB).

Let us think of PPB as embodying *horizontal* comprehensiveness—comparing alternative expenditure packages to decide which of them best contributes to larger programmatic objectives. ZBB, by contrast, might be thought of as manifesting *vertical* comprehensiveness: every year alternative expenditures from base zero are considered for all governmental activities or objectives treated as discrete entities. In a word, *PPB compares programs, ZBB compares alternative funding*.

Planning, Programming and Budgeting

The strength of PPB lies in its emphasis on *policy analysis* to increase effectiveness: programs are evaluated, found wanting, and presumably replaced by alternatives, designed to produce superior results. Unfortunately, PPB engenders a conflict between error *recognition* and error *correction*. There is little point in designing better policies so as to minimize their prospects of implementation. But why should a process devoted to policy *evaluation* end up stultifying policy *executions*? Answer: because PPB's *policy rationality* is countered by its *organizational irrationality*.

If error is to be altered, it must be relatively easy to correct.[6] But PPB makes it hard. The "systems" in PPB are characterized by their proponents as highly differentiated and tightly linked. The rationale for program budgeting lies in its connectedness: like programs are grouped together. Program structures are meant to replace the confused concatenations of line-items with clearly differentiated, non-overlapping boundaries—only one set of programs to a structure. This means that a change in one element or structure must result in change reverberating throughout every element in the same system. Instead of altering only neighbouring units or central control units, which would make change feasible, all are, so to speak, wired together so that the choice is, in effect: all or none.

Imagine one of us deciding whether to buy a tie or a kerchief. A simple task, one might think. Suppose, however, that organizational rules require us

---

[6] This and the next eight paragraphs are taken from my "Policy Analysis is What Information Systems are Not," *New York Affairs* 4: Spring 1977.

to keep our entire wardrobe as a unit. If everything must be rearranged when one item is altered, the probability that we will do anything in those circumstances is low. The more tightly linked and the more highly differentiated the elements concerned, the greater the probability of error (because tolerances are very small), and the less the likelihood that error will in fact be corrected (because with change, every element has to be recalibrated with every other which had been previously adjusted). Being caught between revolution (change in everything) and resignation (change in nothing) has little to recommend it.

Program budgeting increases rather than decreases the cost of correcting error. The great complaint about bureaucracies is their rigidity. As things stand, the object of organizational affection is the bureau as serviced by the unusual line-item categories from which people, money, and facilities flow. Viewed from the standpoint of bureau interests, programs to some extent are negotiable: some can be increased, others decreased, while keeping the agency on an even keel, or, if necessary, adjusting it to less happy times without calling into question its very existence. Line-item budgeting, precisely because its categories (personnel, maintenance, supplies) do not relate directly to programs, are easier to change. Budgeting by programs, precisely because money flows to objectives, makes it difficult to abandon objectives without abandoning the organization which gets its money for them. It is better to use non-programmatic rubrics as formal budget categories, permitting a diversity of analytical perspectives, than to transform a temporary analytic insight into a permanent perspective through which to funnel money.

The good organization is interested in discovering and correcting its own mistakes. The higher the cost of error—not only in terms of money but also in personnel, programs, and prerogatives—the less the chance that anything will actually be done about them. Organizations should be designed, therefore, to make errors visible and correctible, that is to say, cheap and affordable.

Zero-base Budgeting

The ideal a-historical information system is zero-base budgeting. The past, as reflected in the budgetary base (common expectations as to amounts and types of funding), is explicitly rejected: there is no yesterday; nothing is to be taken for granted, everything is at every period subject to searching scrutiny. As a result, calculations become unmanageable. The same is true of PPB, which requires comparisons of all or most programs which might contribute to common objectives. To say that a budgetary process is a-historical is to conclude that it increases the sources of error while decreasing the chances of correcting mistakes: if history is abolished, nothing is settled. Old quarrels become new conflicts. Both calculation and conflict increase exponentially, the former worsening selection, and the latter obstructing correction of error. As the number of independent variables grows, ability to control the future declines (because the past is assumed not to limit the future). As mistrust grows with conflict, willingness to admit, and hence to correct, error diminishes. Doing without history is a little like abolishing memory—momentarily convenient, perhaps, but ultimately embarrassing.

Only poor countries come close to zero-base budgeting, not because they *wish* to do so, but because their uncertain financial position continually forces them to go back on old commitments. Because past disputes are part of present conflicts, their budgets lack predictive value; little that is stated in them is likely to occur. A-historical practices, which are a dire consequence of extreme instability and from which all who experience them devoutly desire to escape, should not be considered normative.

Policy Implications

ZBB and PPB share an emphasis on the virtue of objectives. Program budgeting is all about relating larger to smaller objectives among different programs and zero-base budgeting promises to do the same within a single program. The policy implications of these methods of budgeting, which distinguish them from existing approaches, derive from their overwhelming concern with ranking objectives. Thinking about objectives is one thing, however; making budget categories out of them is quite another. Of course, if one wants the objectives of today to be the objectives of tomorrow, if one wants no change in objectives, then building the budget around objectives is a brilliant idea. But if one desires flexibility in objectives (sometimes known as learning from experience), it must be possible to change them without simultaneously destroying the organization through the withdrawal of financial support.

Both PPB and ZBB are expressions of the prevailing paradigm of rationality in which reason is rendered tantamount to ranking objectives. Alas! An efficient mode of presenting results in research papers—find objectives, order them, choose the highest valued—has been confused with proper processes of social inquiry. For purposes of resource allocation, which is what budgeting is about, ranking objectives without consideration of resources is irrational. The question cannot be "what do you want?"—as if there were no limits—but should be "what do you want compared to what you can get?" Ignoring resources is as bad as neglecting objectives, as if one were not interested in the question "why do I want to do this?" After all, an agency with a billion would not only do more than it would with a million but might well wish to do something quite different. Resources affect objectives as well as vice versa. And budgeting should not separate what reason tells us belongs together.

For purposes of economic management, comprehensive calculations stressing efficiency (ZBB) and effectiveness (PPB) leave much to be desired. For one thing, comprehensiveness takes time, and this is no asset in responding to fast-moving events. For another, devices which stress the intrinsic merits of their methods—"this is (in)efficient and that is (in)effective"—rub raw when good cannot be done for external reasons, that is, because of the state of the economy. Co-operation will inevitably be compromised when virtue in passing one test becomes vice in failing another.

I have already stated that conflict is increased by a-historical methods of budgeting. Here I wish to observe that efforts to reduce conflict only make matters worse by vitiating the essential character of comprehensiveness. The cutting edge of competition among programs lies in postulating a range of policy objectives small enough to be encompassed and large enough to overlap so that there are choices (trade-offs in the jargon of the trade) among them. Instead, PPB generated a tendency either to have only a few objectives, so anything and everything fit under them, or a multitude of objectives, so that each organizational unit had its own home and did not have to compete with any other.[7] ZBB produced these results in this way: since a zero base was too threatening or too absurd, zero moved up until it reached, say 80 per cent of the base. To be sure, the burden of conflict and calculation declined—but so, at the same time, did any real difference from traditional incremental budgeting.

In so far as financial control is concerned, ZBB and PPB raise the question: control over what? Is it control over the *content* of programs? Or the efficiency of a given program? Or the total cost of government? Or just the legality of expenditures? In theory, ZBB would be better for efficiency, PBB for effectiveness, and traditional budgeting for legality. Whether control extends to total costs, however, depends on the form of financing, a matter to which we now turn.

APPROPRIATIONS OR TREASURY BUDGETING

A traditional budget depends on traditional practice—authorization and appropriation followed by expenditure, post-audited by external auditors. But in many countries, traditional budgeting is not in fact the main form of public spending. Nearly half or more of public spending does not take the form of appropriations budgeting but what I shall call "treasury budgeting." I find this nomenclature useful in avoiding the pejorative connotations of what would otherwise be called "back-door" spending, because it bypasses the appropriations committees in favour of automatic disbursement of funds through the treasury.

---

[7] See Jeanne Nienaber and Aaron Wildavsky (1973).

Alternatives to Traditional Appropriations

For our present purposes, the two forms of treasury budgeting which constitute alternatives to traditional appropriations are *tax expenditures* and *mandatory entitlements*. When concessions are granted in the form of tax reductions for home ownership or college tuition or medical expenses, these are equivalent to budgetary expenditures except that the money is deflected at the source. In the United States, tax expenditures now amount to over $100 billion a year. In one sense, this is a way of avoiding budgeting before there is a budget. Whether one accepts this view is a matter of philosophy. It is said, for instance, that the United States government has a progressive income tax. Is that the real tax system? Or is it a would-be progressive tax as modified by innumerable exceptions? The budgetary process is usually described as resource allocation by the president and Congress acting through its appropriations committees. Is that the real budgetary process? Or is it that process together with numerous provisions for "back-door" spending, low interest loans, and other devices? From a behavioural or descriptive point of view, actual practices constitute the real system. Exceptions are part of the rule. Indeed, since less than half of the budget passes through the appropriations committees, the exceptions *must* be greater than the rule. And some would argue that the same could be said about taxation. If the exceptions are part of the rule, however, tax expenditures stand in a better light. Then the government is not contributing or losing income: instead it is legitimately excluding certain private activities from being considered as income. There is no question of equity—people are just disposing of their own income as they see fit in a free society. Unless whatever is is right, however, tax and budget reformers will object to sanctifying regrettable lapses as operating principles. To them, the real systems are the ones which we ought to perfect: a progressive tax on income whose revenues are allocated at the same time through the same public mechanism. And tax expenditures interfere with both these ideals.

Mandatory, open-ended entitlements—our second category of treasury budgeting—provide that anyone eligible for certain benefits must be paid, regardless of the total. Until the legislation is changed or a "cap" limits total expenditure, entitlements constitute obligations of the state through direct drafts on the treasury. Were I asked to give an operational definition of the end of budgeting, I would say "indexed, open-ended entitlements." Budgeting would no longer involve allocation within limited resources but only addition of one entitlement to another, all guarded against fluctuation in prices.

Obviously, treasury budgeting leaves a great deal to be desired in controlling costs of programs, since these depend on such variables as levels of benefits set in prior years, rate of application, and severity of administration. Legal control is possible but difficult because of the large number of individual cases and the innumerable provisions governing eligibility. If the guiding principle is that no one who is eligible should be denied, *at the cost of including some who are ineligible*, expenditures will rise. They will decline if the opposite principle—no ineligibles even if some eligibles suffer—prevails.[8]

Whether or not entitlement programs are efficient or effective, the *budgetary process* will neither add to nor subtract from that result simply because it plays no part. To the extent that efficiency or effectiveness are spurred by the need to convince others to provide funds, such incentives are either much weakened or altogether absent. The political difficulties of reducing benefits or eliminating beneficiaries speak eloquently on this subject. No doubt benefits may be eroded by inflation. Protecting against this possibility is the purpose of indexing benefits against inflation (thus doing for the individual what volume budgeting does for the bureaucracy).

Why, then, in view of its anti-budgetary character, is treasury budgeting so popular? Answer: because of its value in coping with conflict, calculation, and economic management. After a number of entitlements and tax expenditures have been decided upon at different times, usually without full awareness of the others, implicit priorities are produced *ipso facto*, untouched, as it were, by human hands. Conflict is reduced, for the time being at least, because no explicit decisions giving more to this group, and less to another, are necessary. Ultimately, to be sure, resource limits will have to be considered, but even then only a few rather than *all* expenditures will be directly involved, since the others go on, more or less, automatically. Similarly, calculation is contracted as treasury budgeting produces figures, allowing a large part of the budget to be taken for granted. Ultimately, of course, there comes a day of reckoning in the form of a loss of flexibility due to the implicit pre-programming of so large a proportion of available funds. For the moment, however, the attitude appears to be "sufficient unto the day is the (financial) evil thereof."

For the purposes of economic management, treasury budgeting is a mixed bag. It is useful in providing what are called automatic stabilizers. When it is deemed desirable not to make new decisions every time conditions change, for example, regarding unemployment benefits, an entitlement enables funds to flow according to the size of the problem. The difficulty is that not all entitlements are counter-cyclical (child benefits, for example, may rise independently of economic conditions), and the loss in financial flexibility generated by entitlements may hurt when the time comes to do less.

---

[8] The importance of these principles is discussed in my book, *Speaking Truth to Power: The Art and Craft of Policy Analysis* (Boston: Little, Brown, forthcoming).

Importance of Time

Nevertheless, treasury budgeting has one significant advantage over appropriations budgeting, namely *time*. Changes in policy are manifested quickly in changes in spending. In order to bring considerations of economic management to bear on budgeting, these factors must be introduced early in the process of shaping the appropriations budget. Otherwise last-minute changes of large magnitude will cause chaos by unhinging all sorts of prior understandings. Then the money must be voted and preparations made for spending. In the United States the completion of this process—from the spring previews to the Office of Management and Budget, to the president's Budget in January, to congressional action by the following summer and fall, to spending in the winter and spring—takes from 18 to 24 months. This is not control but remote control.

''Fine-tuning expenditures,'' attempting to make small adjustments so as to speed up or slow down the economy, do not work well anywhere. Efforts to increase expenditure are just as likely to decrease it in the short run due to the very effort required to expand operations. Similarly, efforts to reduce spending in the short run are just as likely to increase it due to such factors as severence pay, penalties for breaking contracts, and so on. Hence, even as efforts continue to make expenditures more responsive, the attractiveness of more immediate tax and entitlement increases is apparent.

The recalcitrance of all forms of budgeting to economic management is not so surprising: after all, both spending programs and economic management can not be made more predictable if one is to vary in order to serve the other. In an age profoundly influenced by Keynesian economic doctrines, with their emphasis on the power of government spending, however, continued efforts to link macro-economics to micro-spending are only to be expected.

THE STRUCTURAL BUDGET MARGIN

One such effort is the ''structural budget margin'' developed in the Netherlands. Due to dissatisfaction with the Keynesian approach to economic stabilization, as well as to disillusion with its short-term fine-tuning, the Dutch sought to develop a longer-term relationship between growth of public spending and the size of the national economy. Economic management was to rely less on sudden starts and stops of taxation and expenditure, and greater effort was to be devoted to controlling public spending. The closest the United States has come to something similar is through the doctrine of balancing the budget at the level of full employment, which almost always entails a deficit. The Dutch were particularly interested in a control device because of the difficulty of getting agreement to hold down expenditures in coalition governments. Thus, spending was to be related not to *actual* growth but to *desired* growth, with only the designated margin available for new expenditure.[9]

Needless to say, there are differences in definition of the appropriate structural growth rate, and it has been revised up and down. Since the year which is chosen as a base makes a difference, that too has been in dispute. And, as one would expect, there are disagreements over calculation of cash or volume of services, with rising inflation propelling a move toward cash. Furthermore, since people learn to play any game, *conservative* governments used the structural budget margin to hold down spending while socialists used the mechanism to increase it, for then the margin became a tool for calculating the necessary increases in taxation. Every way one turns, it appears, budgetary devices are good for some purposes and bad for others.

WHY THE TRADITIONAL BUDGET LASTS

Every criticism of traditional budgeting is undoubtedly correct. It *is* incremental rather than comprehensive; it *does* fragment decisions, usually making them piecemeal; it *is* heavily historical, looking backward more than forward; it *is* indifferent about objectives. Why, then, has traditional budgeting lasted so long? Answer: *because it has the virtues of its own defects*.

Traditional budgeting makes calculations easy, precisely because it is not comprehensive. History provides a strong base on which to rest a case. The present is appropriated on the basis of the past, which may be known, rather than of the future, which cannot be comprehended. Choices which might cause conflict are fragmented so that not all difficulties need to be faced at one time. Budgeters may have objectives but the budget itself is organized around activities or functions—personnel, maintenance, and so on. One can change objectives, then, without challenging organizational survival. Traditional budgeting does not demand analysis of policy; neither, however, does it inhibit it. Because it is neutral in respect of policy, traditional budgeting is compatible with a variety of policies, all of which can be converted into line-items. Budgeting for one year at a time has no special virtue (two years, for instance might be just as good or better) except in comparison with more extreme alternatives. Budgeting several times a year aids economic adjustment but also creates chaos in departments, disorders calculations, and worsens conflict. Multi-year budgeting enhances planning at the expense of adjustment, accountability, and possible price stability. Budgeting by volume and entitlement also aids planning and efficiency in government at the cost of control and effectiveness. Budgeting becomes spending. Traditional budgeting lasts, then, because it is simpler, easier,

---

[9] J. Diamond (1977).

more controllable, more flexible than modern alternatives, such as ZBB, PPB, and indexed entitlements.

A final criterion has not been mentioned because it is inherent in the multiplicity of others, namely *adaptability*. To be useful, a budgetary process should perform tolerably well under all conditions. It must perform in face of the unexpected, deficits and surpluses, inflation and deflation, economic growth and economic stagnation. Because budgets are contracts within governments, signifying agreed understandings, and signals outside of government, informing others of what government is likely to do so that they can adapt to it, budgets must be good (though not necessarily excellent) for all seasons. It is not so much the fact that traditional budgeting succeeds brilliantly on every criterion, as that it does not fail entirely on any one, which is responsible for its longevity.

Needless to say, traditional budgeting also has the defects of its virtues: no instrument of policy is equally good for all purposes. Although budgets look back, they may not look back far enough to understand how (or why) they got to where they are. Comparing this year with last may not mean much if the past was a mistake and the future is likely to be a bigger one. Quick calculation may be worse than none if it is grossly in error. There is an incremental road to disaster as well as faster roads to perdition. Simplicity may become simple-mindedness. Policy neutrality may degenerate into lack of interest in programs. Why then has it lasted? Answer: so far, no one has come up with another budgetary procedure which has the virtues of traditional budgeting, while lacking its defects.

At once one is disposed to ask why it is necessary to settle for second or third best: why not combine the best features of the various processes, specially selected to work under prevailing conditions? Why not multi-year volume entitlement for this and annual cash zero-base budgeting for that? The question answers itself; there can only be one budgetary process at a time. It follows that the luxury of picking different ones for different purposes is unobtainable. Again, the necessity of choosing the least bad, or the most widely applicable over the largest number of cases, is made evident.

Still, almost a diametrically opposed conclusion is also obvious to students of budgeting: observation reveals that a number of different processes do in fact co-exist right now. Some programs are single-year while others are multi-year; some have cash limits while others are open-ended or even indexed; some are investigated in increments while others (where repetitive operations are involved) receive, in effect, a zero-base review. Thus beneath the facade of unity, there is in fact diversity.

How, then, are we to choose among truths that are self-evident (there can be only one form of budgeting at a time and there are many)? Both cannot be correct when applied to the same sphere; but, I think both are true when applied to different spheres. The critical difference is between the financial form in which the budget is voted in the legislature, and the different ways of thinking about budgeting. It is possible to analyse expenditures in terms of programs, over long periods of time, and in many other ways, without requiring that the *form of analysis* should be the same as the form of appropriation. Indeed, as we have seen, there are persuasive reasons for insisting that form and function should be different. All this can be summarized: the more neutral the form of presenting appropriations, the easier to translate other changes—in program, direction, organizational structure—into the desired amounts without turning the categories into additional forms of rigidity, acting as barriers to future changes.

Nonetheless, traditional budgeting must be lacking in some respects or it would not be replaced so often by entitlements or multi-year accounts. Put another way, treasury budgeting must reflect strong social forces. These are not mechanisms to control spending but to increase it.

"The Budget" may be annual but tax expenditures and budget entitlements go on until changed. Where there is a will to spend, there is a way to do so.

CONCLUSION

I write about auditing largely in terms of budgeting and budgeting largely in terms of public policy. The rise of big government has necessarily altered our administrative doctrines of first and last things. When government was small, so also was public spending. Affairs of state were treated as extensions of personal integrity . . . or the lack thereof. The question was whether spending was honest. If public spending posed a threat to society, it was this: that private individuals would use it to accumulate private fortunes with which to enter the economy. State audit was about private avarice. As government grew larger, its manipulation meant more. Was it doing what it said it would do with public money? *State audit became state compliance*. But when government became gigantic, the sheer size of the state became overwhelming. The issue was no longer control of the state—getting government to do what it was told—but the ability of the state to control society. Public policy, that is, public measures to control private behaviour, leapt to the fore. And that is how auditing shifted from private corruption to government compliance to public policy.

Social forces ultimately get their way, but while there is a struggle for supremacy, the form of budgeting can make a modest difference. It is difficult to say, for instance, whether the concept of a balanced budget declined due to social pressure or whether the concept of a unified budget, including almost all transactions in and out of the economy, such as trust funds, makes it even less likely. In days of old when cash was cash, and perpetual deficits were not yet invented, a deficit meant more cash out than came in. Today, with a much larger total, estimating plays a much more important part, and it is anyone's guess (within a margin of $50 billion) as to the actual state of affairs. The lesson to be drawn is that for purposes of accountability and control, the simpler the budget the better.

Taking as large a view as I know how, the suitability of a budgetary process under varied conditions depends on how well diverse concerns can be translated into its forms. For sheer transparency, traditional budgeting is hard to beat.

Chapter Four

Program Evaluation in the Federal Government

by
Harry Rogers

INTRODUCTION

It would not be surprising to find some slight scepticism about a new central agency (the Comptroller General's office) related to reforms in management and budgetary techniques. Let me assure you that the guiding philosophy of the Office will be one of practicality and realism. My intention is that the Office of the Comptroller General will be a catalyst for more efficient, effective, and responsive administration of government expenditures, which will enable the federal government to do *more* with its resources. We will not attempt to implement grand and expensive systems thinking that they will provide easy cures to diverse and complex problems. Systems must be tools for some clear purpose or set of purposes. Elegant systems which do not take into account the complexity of the problems and the talents and needs of their operators soon fail.

I lived through the sobering experience that U.S. business had with its experiments in computer modelling of the business environment, in the second half of the 1960s. In good faith, and pushed by the MBA's that were hired from MIT, Wharton, Carnegie-Mellon, Duke and elsewhere, business delivered its third-generation computer power to them and to the corporate vice presidents of planning that sprouted on the organization trees in that period. Business leaders bought the thesis that macro-simulations of the critical choices, required to maintain and improve market share and profit growth, would reduce the risk levels for them and leave even more time for golf.

While there have been some modest successes, this whole effort collapsed under its own weight, for predictable reasons. The efforts were undertaken on too grand a scale, were simply not tied to the main stream of the management of the enterprise, and, for these reasons, were divorced from the needs of operational managers who make the commitments and deliver the results—in other words, divorced from the processes of accountability within the business firm.

Well, so much for systems cures.

THE OFFICE OF THE COMPTROLLER GENERAL

I would like to start by reviewing some facts about the organization and responsibilities of the Office of the Comptroller General, to provide a context for my remarks on program evaluation which will follow. First, I have a direct reporting relationship to the President of the Treasury Board. The Treasury Board, as you know, is the Cabinet's committee on both expenditure control and management of the federal public service. The Board's mandate is implemented through two departments, the Office of the Comptroller General and the Treasury Board Secretariat. The Secretary of the Treasury Board Secretariat, Dr. Maurice LeClair and myself, as Comptroller General, are both senior deputy ministers and have the same responsibilities for policy advice and implementation as other deputy ministers. Our joint task, provided through the authority of the *Financial Administration Act*, is to ensure that the expenditures of the government of Canada are well planned and controlled, and that the federal public service is managed economically, efficiently, and effectively.

The Office of the Comptroller General was created by an Act of Parliament in June 1978. It has broad authority and responsibility for financial management and related administrative practices and controls in the government of Canada. For instance, we establish the accounting principles and practices for the financial statements of the government of Canada and the structure of accounts to be used for the preparation of financial reports by government departments and agencies. We are also vitally concerned with the form and content of financial reports, so that we can improve their use by government departments and agencies in problem identification and decision making. We establish the internal *audit* ground rules on financial matters that are to be followed by departments and agencies. We conduct regular reviews of all these activities with departments to identify the extent of their compliance and the need for them to take corrective actions or make further progress. We play a significant role in the development of training programs for financial officers in government, and in the hiring standards and grades that are established for these positions. We also participate in the selection processes of senior financial officers. We share this role with the hiring departments, the Public Service Commission and the Personnel Policy Branch of the Treasury Board Secretariat.

But the responsibilities of my Office of primary interest in the present context are the responsibilities we have for the development of government-wide procedures to ensure that the efficiency and effectiveness of government operations are measured by departments and agencies. These responsibilities are discharged through two major Treasury Board policies. The first applies to the routine measurement of the performance of government programs. The second has to do with procedures that departments and agencies use to carry out the periodic evaluation of all the programs under their jurisdiction.

In short, my Office is responsible for the systems by which the financial function is administered and controlled, and for the related non-financial planning and operational control activities upon which the exercise of financial control depends. My functions have to do with procedures and processes, or with the "delivery systems" used by departments and agencies in managing their affairs.

If I am primarily concerned with the "how" of the financial function, Dr. LeClair as head of the other body reporting to the president of the Treasury Board, is responsible for the "what" aspects. The Secretariat has responsibility for the direction, planning, review, and control of expenditure plans and programs and their relative priorities, general administrative policy, and for government policies and practices in the fields of personnel management and official languages. In summary terms, the Secretariat is responsible for advising Treasury Board ministers on the selection of programs and projects that will achieve the objectives of the government in the most effective manner and in accordance with its priorities. In addition, the Secretariat has to ensure that the manpower and material resources that are approved are used by departments and agencies for their intended purposes.

Let me put these responsibilities in a broader context. Individual ministers of our program departments and their senior officials are accountable for the actual delivery of goods and services to the public, and for the competence with which the managerial functions are performed within their department. The president of the Treasury Board is accountable for whether or not the federal public service is managing well, as a whole. My Office was established to assist with this endeavour by working for the best realizable processes and procedures to be implemented on a service-wide basis.

THE MEANING OF PROGRAM EVALUATION

Having now covered briefly the responsibilities of the Treasury Board, the Treasury Board Secretariat and the Office of the Comptroller General, I would like to move on to program evaluation.

The term "program evaluation" is now widely used as an umbrella to cover a great variety of attempts to arrive at objective and coherent overviews of the effectiveness with which the tasks of government are performed. Program evaluation is no one technique, but is in effect the application of the best current empirical methods of investigation in the social sciences. The aim is to generate information about whether or not specified goals are being reached well, or at all. But if program evaluation is no one thing, I think it is safe to say that knowing how well effort translates into effect is always the objective. One popular definition states that program evaluation means that one accumulates evidence on the manner and extent to which appropriated funds produce the intended conditions. The complication here is that "intended conditions" can be specified at a number of levels. This brings me to the *two* major uses of the term "program evaluation" in the federal public service. For the purposes of simplicity, let me categorize these as big "P" and small "p" programs. Each has important implications for my Office, in relation to the distinction I made earlier between management within departments and management across the whole public service, management of departments if you like.

The Big "P" Definition of Program Evaluation

The term "program evaluation" has been used to refer to the process of evaluating the large programs of the government of Canada. This has come to be known in my Office as the "big P" definition of program evaluation. Those of you familiar with the form of the planning, programming and budgeting system, or PPBS, introduced in the late 1960s, will know that the word "program" is used in this big "P" sense. For PPBS purposes, a program is simply the entire set of public service activities serving one of the government's major objectives. For example, there is the Defence Program, composed of the entire activities of the Department of National Defence. There is an Agricultural Program, and so on, through a very considerable number of programs, each serving a major governmental objective.

PPBS was created out of a desire to find a way to help vast organizations improve control over their gangling limbs and segmented nervous systems. It was, and is, an ambitious attempt, whose broad aims have met with mixed success.

The components which make up the programming, planning and budgeting system are generally accepted to be:

1. the clear statement of objectives
2. the evaluation of program results as these results relate to objectives
3. the computation of total system costs
4. the assessment of alternative means of providing the same service or one which is an acceptable substitute
5. the integration of policy and program decisions with the budgetary process

The best known evidence of the implementation of PPBS is the Main Estimates of the government of Canada which are tabled for each financial year. Beginning with the Main Estimates for 1970-71, there was a radical change from the form followed up to that time. The change, to aggregating and displaying financial information by functions and programs, in addition to a line-item basis such as salaries, helps to guide thinking to the link between resources and governmental objectives. Think for a moment about the potential power of a summary which can aggregate the activities of about 2,200 sub-programs into the 170 expenditure programs of the Main Estimates, then into the 11 broad policy areas of government. In passing, it would be of interest, I think, for you to know that the Treasury Board Secretariat is currently undertaking a major review of the form and content of the Main Estimates. My Office is working closely with the Secretariat in this review.

Generally, then, PPBS can be viewed as an attempt to organize and push forward the central aims of strategic planning, improvement in managerial efficiency, and the exercise of adequate financial control. Evaluation, in the common-sense meaning of reviews and analyses of results of program decisions, has always been part of the PPB system.

I would judge that it is the PPBS sense of the term "program evaluation" that is of primary interest to Parliament. There was an announcement in the Throne Speech of October 1978 concerning the review by Parliament of evaluations by the government of major programs. The objective is to put in place an explicit legislative base to permit the transmittal to Parliament of a steady flow of program evaluations. Parliament will then be able to assess these big "P" programs by comparing actual achievement with stated objectives. It is still premature to discuss the parliamentary mechanisms which can be put in place to fulfill this function.

I should note that there has always been parliamentary review of government evaluations of major programs through special studies by various committees of the House of Commons and the Senate, various "White" papers, reports of Royal Commissions and so forth.

The Small "p" Definition of Program Evaluation

In the policy statement of the Treasury Board concerning the evaluation of government programs, there is the definition of program evaluation that is used in program budgeting, plus a second use of the term. We call this second definition the small "p" definition. This Treasury Board policy, issued in September of 1977, says that deputy ministers are to ensure that their programs are periodically evaluated—first, in terms of their effectiveness in meeting their objectives, and second, in terms of the efficiency with which they are being administered. There are departmental programs, branch programs, directorate programs, and so on. Each program comprises activities which support stipulated program goals which in turn are linked to broader objectives. The basic task is to ensure that small "p" program goals are clearly stated and evaluable, and that the links to broad objectives are also plausible and capable of assessment.

We have a right to expect that each manager will be concerned with two central questions of the effectiveness of a program. These are, "am I doing the right thing in terms of the program's legislative mandate?" and "to what extent am I actually doing it?" The notion of the mandate is of course central, and it is important to acknowledge that the mandate is not always as clearly specified as bureaucrats would like. Managers must also be concerned with economy and ask of themselves, "are the resources I am applying to the program and its operations entirely appropriate to the program?" and "am I acquiring these resources at minimum cost?" Finally, there remains the vital question of how well these resources are being transformed into the desired outputs, which is the critical issue of operational efficiency. And all this is a roundabout way of saying that we have a right to expect that, in the end, the public service can provide designated services to the public with proper value for money.

Professional management implies managers who are continuously seeking improvements, who are actively pursuing economy in the acquisition of resources; efficiency in the transformation of these resources; and effectiveness in achievement. Professional management implies the existence of the critical managerial skills of expressing seasoned judgment on priorities, the ability to select and motivate high calibre people, and the abilities to organize and to delegate. Further, it implies that managers will and do take whatever steps are necessary to reinforce their managerial judgments based on seasoned experience with hard information, and with rigorous analysis of that information, in order to undertake both short- and long-term action and corrective adjustments. To review: energy, intelligence, commitment, and integrity are not sufficient to guarantee success in the absence of evaluation. An ever-present pitfall for a manager is how easily his raw judgment can be misleading in the absence of hard information. It is fair to say that, without facts, it is not possible to attribute outcomes to managerial performance, or to luck—good or bad—or to a bit of both. Periods of growth cover a multitude of sins and everyone looks good, and times of restraint encourage equally unfounded attributions of blame and waste. This is why, as managers, we need to measure and assess what is happening—to *evaluate*.

In this sense, then, program evaluation in the government of Canada is a necessity at every managerial level. It focuses on economy, efficiency, and effectiveness, separately and in their necessary interrelationships.

WHO REQUIRES PROGRAM EVALUATION?

Let me now address the question of who initiates program evaluation. First, who has historically evaluated government programs when "program" is defined to include only the large aggregations of governmental activities serving the major objectives of the Cabinet and Parliament?

This is, of course, the traditional ongoing concern of Cabinet itself. A major responsibility of the Cabinet in our system is to hold a watching brief for the people, on what is actually going on in the government, and to be alert, therefore, for any signs that its major programs are not being implemented as well as they might. It is the Cabinet which puts the programs in place in the first instance and which has initiated requirements for review.

The Cabinet has three mechanisms for the initiation and conduct of program evaluations. It has a committee system organized in part on substantive policy lines. The Cabinet Committee on Social Policy, for example, might recommend that Cabinet request a review of federal income maintenance programs, as they relate to a particular group in society, the aged. Such evaluations often cut across departmental lines. Secondly, an individual minister accepts the assignment of responsibility for each public service portfolio. In the exercise of his or her responsibility a minister of Energy, Mines and Resources might initiate the review of an energy conservation program. And, finally, there is a detailed assignment of evaluation responsibility to the Treasury Board under the *Financial Administration Act*. As I indicated earlier, this entails both a management dimension to ensure the efficient use of manpower and material resources vis-à-vis all federal departments and agencies, and the responsibility to make day-to-day judgments on the selection of programs and projects consistent with government-wide priorities as determined by Cabinet.

Outside the orbit of the Cabinet there is, of course, the Auditor General's office. The Auditor General now has a broad "value for money" audit mandate. It covers directly the economy and efficiency of government programs, and also permits him, to quote the *Auditor General Act* "to call attention to anything he considers to be of significance and of a nature that should be brought to the attention of the House of Commons, including any cases in which he has observed that . . . satisfactory procedures have not been established to measure and report the effectiveness of programs, where such procedures could appropriately and reasonably be implemented." In terms of major programs, therefore, there are already several evaluation mechanisms in place.

If this represents the superstructure of program evaluation in its proper political context, the underlying responsibility belongs to deputy ministers and heads of agencies. The Treasury Board policy on the Evaluation of Programs by Departments and Agencies, as promulgated in September 1977, clearly establishes that deputy ministers and agency heads, as the chief executive officers of their organizations, are accountable for the effectiveness, efficiency, and economy of all activities, large and small, falling under their direction. The chief executive officers in turn look to each manager reporting to them to assure the quality of those processes and delivery systems falling within the respective individual responsibility centre.

In support of the program evaluation responsibilities of the chief executive officers, a number of specialists and groups have emerged in departments over the past several years, performing operational analysis, performance measurement, internal audit, management audit, and in-depth program evaluation.

My office assumed the functional responsibility for the Treasury Board policy on the evaluation of programs on the creation of the Office in June

1978. We will attend to program effectiveness, efficiency, and economy by assuring that departmental procedures are in place as embodied in the official Performance Measurement and Program Evaluation policies. These are essentially the rules and guide-lines to assist the managers to perform their ongoing management functions described above. In a sense we are the teeth of these rules—or at least the gums.

A BALANCED ASSESSMENT OF THE POTENTIAL FOR PROGRAM EVALUATION

At the outset I emphasized that the guiding philosophy of the Office of the Comptroller General will be practicality and realism. This philosophy is nowhere more appropriate than in the area of program evaluation. The excessive optimism of the sixties about the potential for comprehensive, systematic, integrated program evaluation has given way in some quarters to excessive pessimism as to the ability of program evaluation to improve the quality of government decisions. In a recent publication for the Canadian Tax Foundation, Douglas Hartle (1978), a leading draftsman of the introduction of PPBS to federal government decision making, argues that the techniques are only applicable "to a few programs or policies and a few policy instruments and a few relatively well-established objectives."

I need not dwell on the pitfalls of program evaluation here. The constraints and negative incentives are formidable. Conceptual, analytical, and data problems are potentially severe. A partial list of some of the more obvious pitfalls includes:

- What constitutes a "program" is inherently arbitrary.
- The objectives of a program may not have been sufficiently well formulated to provide a basis for evaluation.
- Many programs have multiple objectives and multiple effects. Quite different weights may be applied to the several objectives at different points in time.
- Appropriate evaluation methodology may not exist.
- The required data may not be readily or economically available.
- The effects of a program may be difficult or impossible to isolate vis-à-vis those of other programs or influences.
- Program managers are vitally involved in seeing that their programs succeed. The incentive is for them to approach evaluation in terms of its potential for program defence and possible expansion, or to resist evaluation because of its potential threat to the continuing existence of the program and to the continuing success of the manager. As Hartle has said, "It is the rare dog that will carry the stick with which it is going to be beaten."

Finally, a particularly important judgmental factor to be taken into account when contemplating the in-depth evaluation of a particular program is the objective assessment of the *cost* of the evaluation relative to the benefits that may result from the information produced. In relation to a program's total budget, there can be no hard and fast rule establishing the percentage of funds that should be allocated to the evaluation function. For instance, major income security programs such as old age security require a relatively *small* percentage of the program's budget for evaluation purposes. On the other hand, relatively large expenditures could be justified for a low-cost program, such as a regulatory program which will have very large or very fundamental impacts on some industry.

In total, these complexities of the effort called program evaluation really indicate that there is no one neat set of techniques to apply across the board. They mean that components of all possible qualitative and quantitative strategies, of judgmental and analytical procedures, must be applied in a creative fashion to problems and activities in context.

It is an immediate priority of my Office to come to grips with this problem by working with departments to establish an agreed, ordered point of departure in the area of program evaluation. For while we bring a new mandate to examine, it is the departments which have the programs and can present us with more or less full accounts of symptoms and reactions to past adjustments. In starting out, however, our strategy has had to take account of an important constraint at both the departmental and central agency level.

This constraint is that there is an evident lack of appropriate, systematic application of the required skills. With this in mind I have recently established a small but important group within my Office, drawn from across the public service, to work initially with the twenty largest departments and then with the *balance* of federal departments and agencies. The group's mandate is to help develop an understanding of the state of the art of program evaluation, as practised in the federal government, in terms of its potential and its limitations, clarifying what the best application of evaluation mechanisms is in the fact of limited resources and skills. My Office is working closely with departments and agencies to put into place a number of management practices. I will touch briefly on these.

First, each department and agency must develop an appropriate comprehensive list of program components. Next, we will make efforts to ensure the co-ordination of evaluation activities in each department and agency. Following on this initiative, a plan will be made which indicates when programs are to be evaluated, normally within a three- to five-year cycle. The priorities in the plan will be based on four criteria: the importance of the program in terms of departmental or ministerial priority, the amenability of each program to evaluation, the size of the program, and the expected cost of the evaluation in relation to the size of the program. Obviously, the plans must be adjusted over time as priorities change, in order to improve the "timeliness" of the evaluations in making decisions.

Also of importance, there will be appropriate documentation of all exclusions from the plan. Obviously not all programs are evaluable in a realistic and practical sense, and it would be wasteful and foolish to pretend otherwise. Indeed, part of our task could well be to ensure that evaluations are performed more selectively. Related to this last point, we will ensure that the terms of reference for program evaluations are developed in such a way as to guarantee that all appropriate areas are reviewed. Reasons for excluding important issues from a specific study will have to be documented by departments.

We must also choose appropriate methodology, and ensure that there is due regard for impartiality and objectivity in the conduct of evaluations.

In order to ensure that evaluation reports have the appropriate impact, there must be suitable reporting relationships to ensure that the deputy minister or chief executive officer can be advised of evaluation results and of the adequacy of the follow-ups to evaluation studies. There must be effective procedures to ensure that program managers take appropriate steps to correct valid problems revealed by evaluation studies.

Finally, there will be an annual updating of the evaluation plans, on a department-by-department basis. The outputs of this endeavour are the plans and timetables agreed upon with each department which will provide a detailed understanding for the extension of good practices and procedures for evaluating the efficiency and effectiveness of programs to all the operations and programs of government where they may appropriately and reasonably be applied. These plans will provide the minimum ''start'' point and, as with all good plans, will change over time as necessary, in a realistic and pragmatic fashion.

I am optimistic at the outset of this joint venture with departments by the enthusiastic co-operation and support I have consistently received from senior public service managers. I have also come to recognize that such managers quite rightly are weary of being told that market forces do not operate to discipline and set limits to the growth of the public sector, in a tone which suggests that this is by design of the public service. The private sector has a ready made, indisputable, and immutable indicator of success—profit and loss. This is a simple evolutionary mechanism, so it is said, to ensure survival of the fittest, in the sense of the most economic product possible.

But a public service which attempts to ride hard on the forces which impinge on the quality of human lives has no such simple and sovereign indicator of success or failure. Money and effort are input, and the outputs are the hundreds of operating schemes for regulating and controlling (for example, prisons and secure asylums), for comforting (foster homes and homes for the retarded), for sustaining (unemployment insurance, welfare, hospitals), and for integrating (adoption agencies, half-way houses, work-training plans) individual citizens into the mainstream of our society.

There are no rules about how these broad goals can best be met, because social scientists have not yet met their goals. That is, there is no adequate science which can tell us how to predict or manipulate human behaviour. Human affairs are matters of "more-or-less," matters of opinion. Yet opinion can be informed. There are better and worse gradations of "muddling through." We try to reduce uncertainty by assessing results, weighing and considering alternatives. We have to have a more systematic way of truly testing the limits. We must, in a way, build *in* our own imprecise, quantitative, and qualitative criteria for ensuring survival of the fittest of our imperfect schemes. And we have to conclude that we have not done as much as we could in this area, however difficult and complex. We need more focus in order to arrive at more publicly credible conclusions on what it is reasonable to provide. The drive to evaluation, then, is an attempt to exert more knowledgeable control over the haphazard propagation of programs.

CONCLUSION

I want to close by repeating that neither myself nor my officers underestimate the difficult challenge of carrying out program evaluations and integrating them meaningfully in the management process. We know the serious limitations of evaluation techniques and that we will never be able to (or seek to) provide quantitative answers to many questions which must remain matters of judgment. But neither should we be blinded by prejudice against new and invigorated attempts at program evaluation as the result of the partial failure of both PPBS and the evaluation techniques to date. We must learn from our mistakes. We must push the art as far as we can.

I cannot emphasize too strongly that evaluations are essential if we are to respond effectively to the challenge of restraint, and to meet the pressing need for greater flexibility in redirecting financial and manpower resources within reduced expenditure ceilings. I am convinced that a hard-headed, non-technocratic, step-by-step approach to program evaluation can succeed in bringing to bear the type of objective information as to the effectiveness of programs that is so necessary to the management and control of government expenditures. For all of the inherent problems, program evaluations offer the potential for a rational basis for change.

Chapter Five

The Public Monitoring of Public Expenditure

by
Harold Renouf

My words fall,
Arousing inquisitiveness
Hoping to stir
Different opinions.[1]

The public evaluation of government spending is a vital and challenging issue in today's world, as several events in 1978 attest. Not only has Proposition 13 passed into our language as a meaningful term, but recent announcements by the government of Canada[2] were sharp reminders that government spending and public attitudes towards it are now urgent and inseparable considerations.

We live in a complex interrelated society wherein the rate of change has been almost exponential and, as Alvin Toffler has shown us, where change itself can also lead to disturbing shocks. In this same time period we have witnessed a marked increase in the size and complexity of the governments of the Western world—governments that seem to grow more remote from the people they serve as they increase in size and influence.

At the same time, I sense a desire on the part of the people for a reassertion of individualism. I read this sign in the fear that people have of big government, in their concern for the environment, their reluctance to finance the cost of ever-expanding programs without considering whether the program objective is being met or whether we are getting value for money. If this reassertion of individualism continues, then the citizens will need more and better information by which they can, as individuals and collective groups, take the decisions which are necessary to achieve their objectives and fulfill their responsibilities.

---

[1] *Poems of Rita Joe* (Halifax: Abanaki Press, 1978).

[2] *Editor's Note*. In August 1978 the government announced spending cutbacks of $2.0 to $2.5 billion.

In these circumstances it will not be enough for the parliamentarians to rely entirely upon the philosophies of Edmund Burke and ignore the wishes of an aroused and enlightened citizenry. Nor will it be sufficient for the citizens to continue to look at their governments on a "we-they" basis as if "they" were something quite apart from the citizens who elected them. So the theme—the public evaluation of public expenditures—is timely and the need for our attending to it is clear. However, its articulation is still somewhat cloudy.

It is in that spirit, the spirit of an attempt to contribute towards illumination of the issue, that I am approaching my task. Santayana has reminded us that those who cannot remember the past are condemned to repeat it. What is not so well recognized, perhaps, is that those who would hope to throw fresh light on the present must venture beyond their usual boundaries. It will not do simply to repeat our conventional wisdom.

In times of change the concepts we know and have accepted may continue to appeal to us and influence our thoughts, but fresh concepts are also required. It was Alfred North Whitehead who noted that the art of free society consists first in the maintenance of the symbolic code; and second, in fearlessness of revision, to make certain that the code serves those purposes which satisfy an enlightened reason.

The following discussion relates to concerns which I can articulate, but which I cannot quantify nor document, nor yet contain in existing professional concepts. This being so, perhaps my comments might best be regarded as a venture in citizenship over and above my other interests and responsibilities. No apology to professionalism is needed in this context.

I begin with a notion, which is neither entirely old nor entirely new. It concerns the relationship between the public and governments. We the public—the Canadian community—have responsibility for the direction of our governments, and for the quality of their management. To fulfill this responsibility, we must provide ourselves with appropriate tools.

Please note what I have done. Rather than thinking of the public as simply a mass of individuals, needing to be informed by its governments and educated by its professionals, I have hypothesized the presence of a thinking and responsible entity, the Canadian community. The rights of this community do not derive from its governments, rather it is government that is derived from the community, and for which the community is responsible.

Seen from this perspective, I am persuaded that the theme takes on new challenge. When the term "the public" means "the Canadian community," and when the Canadian community is conceived to be responsible, as shareholders, for giving direction to the management of its institutions, and for reshaping them as necessary, then the focus shifts. The methods and forums that are needed for the public evaluation of public spending are seen to be rather different from those we have been moving towards as we approach the matter from more narrowly framed professional stances.

For example, it seems to me that much that we have thought of as an extension of evaluation procedures for public purposes has been merely their deepening in a narrow course. We have, for instance, been attempting to deepen our capacity for "Ottawa to evaluate Ottawa" rather than for the Canadian community to evaluate its federal government. We have been assisting the bureaucracy to evaluate its own efforts, or Parliament to monitor spending, rather than enhancing the evaluating capacities of the community as a whole with respect to government. And yet it is with the public as a whole—the Canadian community—that overall responsibility for its governing processes may rest.

In the process of assisting our legislators and the bureaucracy to evaluate their own efforts, our evaluation techniques have become refined, but against narrow, rigid, and quite possibly biased criteria. There is evidence, for example, that many Canadians find some satisfaction in their work, and yet we define all payments to labour as costs in our cost-benefit analyses. There is evidence that many of the goods that Canadians consume are not discretionary purchases, and yet we tend to count all potential consumption as benefit.

Again, in the process of assisting our legislators and the bureaucracy to evaluate their own efforts, the structures of our governments have become expanded and elaborated, but more by inadvertence than in response to clear direction. Any Canadian who wishes to discover even the financial scale of his governments can do so only with the greatest difficulty. There are few satisfactory forums in which the community can readily discuss this matter and come to a mature view, let alone convey this view to those to whom we have delegated some of our managerial responsibilities—our governments through our elected representatives. And these representatives in turn seem to have been having difficulty in fulfilling their responsibilities.

It seems to me that we are now reaping the consequences of our failure to sustain or develop the tools for our community's direction of its governments, including our community's evaluation of the spending of its governments. The foundations of these governments, in the consent of the people and the respect and legitimacy the community accords them, appear to have weakened. It may not be accidental that big government has recently emerged in the opinion of many as public enemy number one, ahead of big labour or big business, nor that there is discontent with the level of government spending. These phenomena may simply reflect our growing frustration and perplexity about what to do with these governments that we have created and for which we are responsible.

It is to such a notion, that is to say, the direction by the Canadian community of its governments, that I think we may need to turn at this juncture in our affairs. We need to turn to such a concept both to ensure the continuation of responsible parliamentary government and the control by the public of the public purse. For if the public does not get the chance to outline its dissatisfaction and concerns and the opportunity to bring about necessary changes, then the revolution of a Proposition 13 becomes inevitable.

In this perspective, the public evaluation of public spending emerges as but one aspect of the task which needs accomplishment, albeit a very necessary aspect. The whole question of government is under discussion in the Canadian community. This discussion embraces the role and functioning of the existing economic structure and of alternatives to it. It also embraces the degree to which the community wishes to rely upon formal economic mechanisms and the public sector to accomplish its objectives, and the degree to which it wishes to use other mechanisms. The present discussion concerning voluntarism forms a good example.

Many strands of thought emerge then from this notion of the Canadian community as the proprietor and director of its organizational arrangements. One strand of thought concerns itself with issues of political process, such as constitution making and the reform of parliaments. But that is not the strand which is of direct concern to us here. Our concern is with economic processes, with how the community may go about managing its economic processes, and with the uses which the community wishes to make of its governments in these processes. But whether constitutional or economic, there needs to be dialogue between governments and their electors before understanding can be reached and meaningful progress achieved.

Perhaps it is time to pause here for a moment and put the matter we are discussing in historical perspective. The fact that we more often think of the responsibility for directing economic process as residing in governments themselves may reflect how easily we forget our past. In the past, governments played a much smaller role in our economic affairs, and the community's responsibility for its affairs was much more taken for granted than it is today. Even though it was expressed in an excessive individualism, the responsibility of the community and the relative importance of the government was taken for granted. Today we seem to be returning to or re-inventing such a notion of community, although the style and tenor of the shift so far leaves much to be desired. Today we generally lack adequate information about the precise role which our governments have been playing. We lack community forums such as were more naturally present when our numbers were smaller and our economic interrelationships less complex (both among ourselves and with other countries). We lack meaningful communication with our governments and are experiencing a rising measure of arbitrariness in what is being done for us by our governments. Viewed in this light, the response of the public, because of lack of understanding, is just as arbitrary, although frequently mistaken for apathy. Further, those of us who have responsibilities within these institutions are sometimes receiving only confused direction from the community we serve.

From virtually every perspective then the situation is frustrating, and yet none of us alone and no specific group seems to be to blame. In the words of

Pogo of comic strip fame, "we have seen the enemy and the enemy is us." The simply truth is that we have not been developing our institutional forms and processes to let us keep sensible pace with the changes which have been occurring. As a community we have not, for example, been developing our methods and forums for the evaluation of government activities as rapidly as we should. Nor have we paid anything like sufficient attention to ensure that our considered judgments are known and our wishes carried out. Perhaps even more importantly, we have been trying to delegate too much managerial responsibility to our governments and other formal organizations, and doing too little among ourselves in informal and voluntary ways. The recognition and acknowledgement of the Canadian community as being the responsible entity, and the improvement of its tools of management for the direction of its affairs, may thus be of some urgency.

The notion that the community directs its governments through the political process is almost unexceptionable in a country such as Canada. The parallel notion with respect to the economic process, however, is far more likely to be inverted, with the direction of the economy appearing to emanate from government itself. As well, government is frequently seen as being in the situation of self-management of its own economic activity. It is this notion which I believe is giving us trouble. It results in such phenomena as persons looking to government for solutions to meet problems: a great search for specific loci of blame with respect to inflation; and even of contention about the release of information on government's own evaluation of government programs.

We may be able to unravel our tangled affairs and see which way to proceed only through such a conceptual shift as is inherent in the reassertion of the community's responsibility for its economic affairs. Within this concept I include the uses to which the community puts its various governments and the direction the community gives its corporations, unions, and other entities. Failing that, I cannot foresee anything but a deepening of our difficulties in the economic sphere, and a continued entanglement of our reasoning around such notions as "the public," "public affairs," and "the public sector." This is to say nothing of our own depreciation of ourselves as being merely members of the community—recipients of action, rather than actors ourselves.

Common sense suggests that we are all interdependent and all responsible for our state of affairs, but our present conceptual and institutional frameworks allow us all to stand aside and point the finger at the other. Our role and voice as citizens have virtually disappeared, and our public affairs have come to be dominated by the voices of competing interests. We convene but rarely as citizens to discuss our situation, but we meet daily as specialists to work on narrow fronts. Our media are filled with the blinkered viewpoints of one specialist after another, but it is usually only in letters to the editor that one may find the common sense of citizens.

Our common interest is the victim, and we find ourselves paying heavily for economic activity that should not have to be pursued at all. There can be no question now that much of our economic life, and much of government activity, is directed towards functions that are simply regulatory or remedial. Such activities are the consequence of situations generated by prior activities which have had adverse spill-over effects—a result of having been pursued too narrowly or too far in the first place. The tendency to divide and subdivide responsibilities has left many of us working far from where the consequences of our work are felt. Thus a "we-they" relationship develops respecting problem areas which are in fact common and require joint solution.

A recognition of this pattern, and a gathering together of the threads of responsibility so that the overall pattern may be considered by the community, seems highly desirable. I believe it was Adam Smith who was concerned that specialization might make people irresponsible. If Canadians are not to suffer further the consequences of the great fragmentation and decline of responsibility that has taken place, but are to begin to see more clearly and are to be able to discuss the overall related pattern, then methods and forums which allow this are needed.

The notion which I offer is that we vest the words "the public" with the meaning of the Canadian community that is responsible for directing its political economy. I further suggest we examine the condition of the present methods and forums for pursuing this task, and in particular the methods and forums for considering public expenditure.

Now I do not think it is beyond the realm of possibility that, in a community that was well organized to direct its affairs, the concept of public expenditure would be transmuted into several different concepts. By itself the concept of public expenditure is an abstraction that perhaps only an economist would focus upon. To the untrained and perhaps more integrative mind, government expenditure must reflect two very different kinds of transactions, and lumping them can be as strong a calculation as adding bananas and oranges to get bananaoranges. Some government spending clearly reflects a real use of our scare resources. Other government spending reflects merely our use of governments as a convenient conduit for movements of monies among ourselves. And yet time and again this distinction is not made in government accounting processes, to the confusion of the community and, it seems, of our governments themselves. That we use our governments as conduits is often a matter merely of administrative convenience, having little to do with the nature of the object of expenditure or its level.

Indeed, government spending and the size of government may no longer be positively related. Even the divisions between public and private sector activities, in a world where governments purchase and subcontract extensively, may not be meaningful, for example, construction. A community evaluation of public expenditure could thus become a new search for meaningful distinctions, for meaningful concepts of what constitutes and defines a critical mass of economic activity. Thus launched, fresh and common sense questions become possible.

The history of the extension of government responsibility suggests that much of our turning to government in the past generation has been as last resort, an expression of despair almost that other community mechanisms would not work—mechanisms of forbearance, of self-regulation, of charity, of stewardship, of informal income redistribution, of risk taking, of investment. And this may have been so. But it soon came to pass that we began to turn to governments as a first resort. In the very act of turning to government, a further erosion of community mechanisms seems to have taken place, leading to a further growth of government.

Any of us, faced as members of the community with the question of how we would go about evaluating government spending, what methods and forums are needed, must surely have to respond that there are a number of matters to be discussed. But at the present time the public evaluation of public spending seems to be one that arises from a narrowly framed professional perspective and not the public's sense of priorities. In the public's sense of priorities, the public evaluation of public expenditures would become simply a proxy for an imminent discussion about the whole question of governance and its modes.

The emergence of new questions thus becomes the significant matter. We should define from a community perspective the public evaluation of public expenditures. Should we be hesitant, we are reminded by Santayana about repeating the past, by Whitehead about the goal of an enlightened reason and the need for fearlessness in revision.

I want to return now to the simple notion of a thinking and responsible community giving direction to the management of its institutions. I have developed at some length what seem to me to be its conceptual and general implications. I want now to go back to the notion and move off from it again in order to develop some of its particular and specific implications. To do this I shall offer three illustrations, each drawn from different fields of activity. These illustrations will, I hope, force upon us an understanding of the matter that may not be clear from the conceptual discussion alone. The notion then can stand or fall upon our rejection or acceptance of both its general and specific implications.

My first illustration concerns the responsibilities of the public sector in a Canadian community framework. I do not think that there can be sensible direction by the community of its affairs in the absence of readily available and adequate forecasts of government activity. Such a matter assessed against the notion we are using, clearly assumes an importance that it has lacked, as the absence of such forecasts in today's perceptual climate attests. In the United States and Great Britain, for example, the publication of five-year forecasts of government expenditures is required by law. Why should not Canada adopt this method of information transmission for the evaluation of existing programs and administration and for better corporate and community planning?

My second illustration concerns the responsibilities of the private sector, and I take for my example my own profession, the accounting profession. I believe the accounting profession has particular responsibilities to the community. I believe it has responsibilities for monitoring particular public accounts from time to time and giving the community an independent assessment of their adequacy for its purposes. It also has the responsibility to serve as a resource to the community to assist in the interpretation of public accounts. It is essential to know that government accounts have been audited and the extent to which financial accountability, compliance, and performance have been met. It is equally essential in the new era of community responsibility to put much more emphasis on the need for generally accepted governmental accounting principles and reporting standards to provide the users of governmental reports with sufficient information to make their economic decisions and to monitor public expenditures.

This means that my profession has a special responsibility to encourage the adoption of such standards of accounting and reporting because the stewardship responsibility of governments requires that there be an information flow which is adequate, comparable, and timely to permit the various "publics" within society the opportunity to judge the quality of administration of their governments, and also to provide users with useful information with which to make economic and civic decisions. As Canadians living in a federal state, we typically have not developed instruments of this kind, and yet they are needed for us to give adequate direction to our governmental entities.

My third example concerns another segment of the private sector, sometimes referred to as the fourth estate. The notion of a responsible community, equipped with the tools to direct governments, places heavy responsibilities on our media. In very many instances the media serve as the only conveyor of directions from the community to its governments and of information from governments to the community. An adequate understanding by the citizenry of the pattern of government activity is essential to the process of providing governments with responsible direction. To the present somewhat capricious coverage of government that is dictated by someone's assessment of newsworthiness on a day-to-day basis, there needs to be added the possibility of a more sustained and comprehensive commentary, a two-way flow of dialogue and information. This will require less accent on "instancy" and more upon analysis and reflection.

Now I hope these three illustrations will help to focus our understanding of what is put in issue by the notion I have offered. It may suggest the need, although I have not chosen to develop this theme, for an initial broad public discussion of our governments and the role we wish them to play. Such a discussion would need to be supported by appropriate documentation of their present role. The government of Canada has, of course, already called for such a public dialogue, and taken steps to encourage it through publication of a discussion paper entitled "The Way Ahead." Conversations among Canadians seem now to be clearly needed.

So where do we stand? We have a notion, and a strong suspicion that if we were to take it seriously its flow would take us in directions that are other than those we must expect. Must we choose? Must we reconsider the topics we want to address if we choose to pursue the logic of community responsibility for government expenditure? I think not, but equally there is no question but that, in the frame of reference I have offered, there dwells the possibility of a change in some of the directions of evaluation that we have been pursuing recently in and around our governments. At the very least, it seems to me that there may be a substantial unwillingness of the community to continue the adequate funding of such evaluation exercises that do not seem fairly directly relevant to them.

Rather than abandon the notion in favour of asserting the sustained relevance of our accustomed approaches, however, I want to leave it because I believe I may have done nothing more than delineate a changing shape of public opinion and perception that is already well underway, and that is likely to form the dominant public reality in the near future. However, even if this is not to be so, I think that in any event the notion might be used as a test bed for our conclusions emanating from our present approaches, and also as a device to make their assumptions about the relationship between the public and governments quite visible.

Chapter Six

Parliament and Expenditure Scrutiny and Evaluation

by
Michael English

THE POWER TO TAX

Though it does not appear in the title of this paper, the principle that only Parliament can impose taxation is at the root of parliamentary control of expenditure. If there were no control of expenditure at all (and—as is shown below—the British Parliament has come very near to dropping all detailed control of expenditure), the level of taxation would still impose an ultimate limit upon the level of expenditure. Of course, in a given year or years, the level of expenditure may exceed the level of taxation and borrowing may fill the gap. A state, unlike a person, may borrow from the banking system—"print money" in common parlance—as was done in Britain prior to 1974, but the relationship of this to price inflation has since then been recognized in Britain as elsewhere, so that the rate of inflation in Britain is now, if anything, rather below the average for developed countries. Inflation is not wholly unpopular; if it were it would not have become a phenomenon of all post–Second World War developed societies, but excessive inflation is still wildly unpopular, with the result that most borrowing must come from non-bank public and must be paid for by increasing public sector interest payments, which itself tends to increase public expenditure. If an economy is expanding, such an increase in public expenditure can be borne without an increase in taxation but in Britain, although the economy has expanded, it has tended not to do so as fast as public expenditure has increased. In such circumstances the parliamentary power to tax, which implies the power to resist taxation, can be an effective brake on the total of public expenditure though not, of course, upon its detailed content.

This power to tax is effectively used by Parliament, probably more effectively than Parliament uses its powers to control expenditure as such. The presentation of his Budget by a Chancellor of the Exchequer is still the highlight of each British parliamentary year and its proposed tax changes are discussed for a week on the floor of the House. They are then embodied in an annual Finance Bill discussed on the floor of the House and, at great length, in committee. This procedure can be criticized and has been, by the Expenditure Committee amongst others, but it cannot be disputed that it is of some effectiveness, as was exemplified by this year's (1978) substantial government defeat in the Finance Bill committee. That defeat did not occur incidentally because the present government is a minority government (although that is the case) but because two government (Labour) supporters voted with the opposition. Over the last ten years or so cross-voting of this sort on various subjects has become more and more usual, in parliaments with sizeable government majorities (1966-1970, 1970-1974) as well as in others (1964-1966, 1974–). This probably reflects the greater volatility of the electorate but, whatever its cause, it inevitably leads to greater parliamentary control of the executive and, due to the British procedural rule that only ministers can propose increases in taxation or expenditure, any government defeat on a procedurally permissible motion tends to result in a reduction in taxation.

THE POWER TO IMPOSE A CHARGE

The rule just mentioned is the second parliamentary control over expenditure, if one regards the power to tax as the first. It provides that only a minister of the Crown may propose "a charge upon the people" which means not only an increase in taxation but an increase in expenditure, which might lead to an increase in taxation. This is not a rule invented by the executive to limit the power of the House of Commons, but a true parliamentary control self-imposed by the House of Commons upon itself by its own standing orders. They go back to the period after the Restoration of 1660 and were a reaction from relatively heavy expenditure and taxation during the civil wars, Commonwealth and Protectorate from 1642 to 1660. With one exception, there is hardly any pressure to change this rule which has considerable advantages. It means that non-ministers advocating increased expenditure cannot irresponsibly require it by law without regard to the taxation consequences; instead they advocate it to spending departments of the executive (who often, of course, agree) which in turn must advocate it to the Treasury (which is less likely to agree) and in Cabinet. The procedure of impoundment which the U.S. Congress has had to construct in order to try to force expenditure on the executive simply does not exist nor need to exist in the United Kingdom.

There is one, relatively minor, exception to the lack of desire to change this apparently restrictive rule. Until now it has also applied to expenditure by the House of Commons upon its own staff and services and there was a feeling that this led to undue executive control over the resources available to non-ministerial M.P.'s. The *House of Commons Administration Act*, just passed, transfers the power to propose expenditure on the House itself from the government to a commission of M.P.'s which will only contain one minister.

THE POWER TO APPROPRIATE

Detailed control of expenditure rests on the power to appropriate expenditure, the origin of which is somewhat obscure. Parliament certainly tried to appropriate expenditure as far back as the middle ages but was commonly ignored by the Crown, as it had to be so long as parliaments were intermittent bodies. The effective power to appropriate seems to have resulted from the seventeenth century success of the House of Commons over the Crown. At any rate, in modern theory, expenditure must be appropriated, that is, detailed in an annual *Appropriation Act*. There are certain well-recognized exceptions to this, such as the payment of judges' salaries without annual appropriation, to secure their independence. A glaring modern exception, criticized in the latest Expenditure Committee report (*Financial Accountability to Parliament*, H.C. 661 Session 1977-78), is the Contingencies Fund, fixed in the last century at £120,000 and raised in 1921 to £1,500,000, after a temporary increase to a much larger figure during and just after the First World War. A similar temporary increase after the Second World War has not only continued to the present time but, in 1974, was given a variable limit of 2 per cent of supply (i.e., roughly, of appropriated expenditure). This is currently £826 million, far more than is needed for unforeseen contingencies, the original purpose of the Fund. More than even this limit can be spent because the Fund is repaid from ordinary Supply expenditure and, to the extent that it is, can spend more than its limit. A fund originally intended for genuine contingencies, such as natural disasters, has in fact come to have a much more extensive use, even being used to anticipate expenditure under new legislation (i.e., entirely foreseen) which has only received a Second Reading in the House of Commons.

An even more serious criticism of present procedure is that the annual *Appropriation Act* is never discussed in detail by the House of Commons. Its Second Reading is used by back-benchers to raise grievances and secure ministerial replies to them, but then its committee stage is taken almost formally on the floor of the House. The theoretical reason for this is that the Act, as a Bill, is founded upon Supply Resolutions which have directly been approved by the House. This is technically true and had they been discussed in detail (as no doubt they once were), there would be merit in this procedural rule. The snag, as described below, is that such discussion hardly ever happens nowadays.

THE GRANT OF SUPPLY

Since the *Appropriation Act* is annual, something must be done to anticipate it and what is done is the passage by the House of Commons of Supply Resolutions which approve Supply Estimates presented by the Crown. In themselves such resolutions would have no legal force but, from time to time, the total expenditure they embody is included in a *Consolidated*

Fund Act. Unlike the *Appropriation Act*, *Consolidated Fund Acts* authorize the release of money but do not appropriate it in any detail. They are in fact temporary acts eventually repealed by the next *Appropriation Act*. Meanwhile, however, the *Public Accounts and Charges Act*, 1891, in effect makes the relevant Supply Resolutions temporary appropriations from the time a *Consolidated Fund Act* is passed. *Consolidated Fund Acts*, however, suffer from exactly the same procedural limitations on their discussion as *Appropriation Acts*, so in the end all detailed consideration of expenditure depends on how adequately the Supply Resolutions are discussed.

The problem is that Supply Resolutions, which could be discussed, are hardly ever actually discussed. Supply days have come to be opposition time and this principle is sacrosanct for obvious democratic reasons. Influenced by the modern communications milieu, the opposition uses them for major debates on topical, newsworthy subjects the government is not as keen to have publicized. To use them to discuss arid Supply Resolutions permitting expenditure which, quite possibly, everyone agrees with in principle, is not thought these days to serve a useful political purpose. Hence, the opposition uses its Supply days to discuss motions it has formulated which argue its political case or simply for a wide-ranging criticism of the government on a motion to adjourn the House. The Supply Resolutions still have to be passed by certain prescribed dates but they are usually passed ''on the nod'' because the time available to discuss them has been used up for these other purposes. The important point to notice is that the leadership of the opposition decides on the use of Supply days. Governments do not, government back-benchers do not, and even opposition back-benchers do not so that once an opposition leadership has decided not to discuss a Supply Resolution it cannot be discussed.

COMMITTEE PROCEDURE

None of this would matter very much if there was an adequate committee procedure for the discussion of Supply. If an Appropriation Bill or a Consolidated Fund Bill were discussed in committee as a Finance Bill, the problems outlined above would be solved. This is considerably more, however, than a mere matter of procedure. A general criticism which the Expenditure Committee has levelled at the whole British system is that taxation and expenditure are not discussed together; instead taxation is thoroughly discussed, expenditure is hardly discussed at all, and the combination of the two is only broadly considered when taxation is being discussed. The Treasury accepted an Expenditure Committee recommendation to study the practice of other countries in this respect but so far this has resulted in no change in this basic defect of the system.

Accepting, for the moment, the system as it is, details of expenditure clearly cannot be discussed save in committee. This has been obvious for a very long time and the First World War caused the setting up of an Estimates Committee. This had two main defects. It was a single committee with a vast field to cover and it was not allowed to discuss "policy," that is, it could say that X could be done for less money or needed more money but it could not say that X was itself a waste of time and money and should be abolished or changed to Y. There were many sterile arguments over what was or was not "policy" and, of course, many able M.P.'s were deterred by the whole procedure from serving on the committee.

The result, in 1971, was that the Estimates Committee was replaced by an Expenditure Committee and with the change of name went other procedural changes. The subjects discussed and the recommendations made have ceased to have serious limitation and the arguments about "policy" are dead. The Committee divides itself up into six sub-committees which between them cover the various departments and in which most of the Committee's work is done. A further change, not solely related to the Expenditure Committee but which has proved of great value to it, was a procedural change making it easier for the Committee to sit in public, which the Expenditure Sub-Committees now usually do when hearing evidence.

This was a considerable improvement but there are now suggestions for going further. This year (H.C. 588-1, Session 1977-78) a Procedure Committee (which included several members of the Expenditure Committee) recommended the replacement of the Expenditure and various other committees by a complete set of departmentally related committees, a suggestion which the Expenditure Committee itself originated. The U.K. House of Commons would thus, it is proposed, come to be more similar to North American and European legislatures in having such a committee structure. The prime object of this would be to enable even closer scrutiny of departmental actions and expenditures.

The Procedure Committee did not quite go as far as it could have gone. It was unanimous on the above recommendation but it was divided on the question of how far to give the committees it proposed any delegated power to deny an appropriation. The minority proposal in this respect is given in an Appendix to this paper. The division of opinion was not on party lines but the conservatism of the Procedure Committee was strongly influenced by the fear of some government supporters that bipartisan committees with such a power might limit the freedom of action of what, in the United Kingdom, are usually single-party governments.

THE ESTIMATES AND ACCOUNTS

The Estimates presented annually to Parliament and the Accounts of money actually spent (which are naturally related in form) may be satisfactory to the executive but are not so useful for parliamentary purposes. Over the years their Votes (the items upon which parliamentary approval of excess expenditure is required) have diminished in number whilst total expenditure has, of course, greatly increased. The average vote is now 92 times larger, in real terms, than it was just over a century ago. The number of subheads (for which Treasury but not parliamentary approval of an excess is required) may have increased, but opportunities for parliamentary control have declined. Alternations in this and other defects of the Estimates and Accounts have just been suggested by the Expenditure Committee in their report *Financial Accountability to Parliament* (H.C. 661, session 1977-78) mentioned above.

AUDIT

The final aspect of parliamentary control of expenditure is the audit of past expenditure. The present basis, under the *Exchequer and Audit Departments (E. & A.D.) Act*, 1866 and 1921, is that the auditor is a Comptroller and Auditor General (C. & A.G.) who has the independence of tenure of a judge (i.e., he can only be dismissed by the Crown at the request of both Houses of Parliament), and is the head of an Exchequer and Audit Department. He reports, at his discretion, to a House of Commons Committee on Public Accounts. Until recently this Committee, of its own choice, always met in private and it takes evidence only from civil servants, never from ministers.

Severe criticisms have been levelled at the Exchequer and Audit Department by both the Expenditure Committee and the Procedure Committee and by outside sources, whilst the financial probity of one organization subject to E. & A.D. audit is currently the subject of investigation by a tribunal of enquiry. This enquiry is also charged with investigating the efficiency of the relevant audit. The expenditure Committee's main recommendations, in its report on the Civil Service (H.C. 535, session 1976-77) were:

(a) The E. & A.D. Act should be amended and should state as principle that the E. & A.D. may audit any accounts into which public money goes even if such public money is not the bulk of receipts into such accounts. Where public money is the bulk of receipts into an account, the E. & A.D. should always audit them, subject only to such specific exceptions as are made in the amended Act. At present the Treasury may direct the E. & A.D. to audit anything but the E. & A.D. may not do so of its own volition. Local governments are subject to audit by auditors who are civil servants in a central government department and nationalized industries and many other quasi-autonomous organizations are audited by commercial auditors chosen by themselves.

(b) The E. & A.D. should be empowered to conduct audits of the management efficiency and effectiveness of all that it audits financially.

(c) The E. & A.D. should change its recruitment policy still further, to provide staff capable of conducting extended audits of the kind mentioned above. Until 1975 the E. & A.D. did not (unlike the rest of the civil service) recruit graduates or professional accountants.

(d) Future C. & A.G.'s should only be appointed after the Committee on Public Accounts has been consulted. Until now each C. & A.G. has been a high Treasury civil servant.

(e) The E. & A.D. staff should be placed under the new House of Commons Commission. At present they are civil servants of the Crown.

The government reply to these recommendations in effect rejected most of them and they are also being resisted by the local authority associations and the professional bodies of the accountancy profession. The government did, however, accept the principle of consultation before the prime minister appoints the next Comptroller and Auditor General. The Expenditure Committee, naturally enough, is not satisfied with this nor is the Procedure Committee, and the matter is now one which will have to be discussed on the floor of the House in the new, 1978-79 session.

CONCLUSION

The conclusion must be that in the United Kingdom parliamentary control of expenditure is very weak but it is not quite as weak as it was before the Expenditure Committee was set up in 1971. The proposals of that Committee and the Procedure Committee would greatly strengthen it but, if past reactions to similar proposals are any guide, many will be resisted by the government and possibly by the opposition front bench as well. The outcome, which the House of Commons itself must decide, will probably be some increase in parliamentary control, though not as much as some of us would like.

It is, of course, possible that a democratic legislature is not the best means of controlling public expenditure. It is not some deep-laid plotting by the executive that has made House of Commons control weak but decisions of the House itself, which may (or may not) be reversed in the future. In the Expenditure Committee, for example, those of its sub-committees which relate to spending departments tend to make suggestions which would increase expenditure rather than reduce it. The reluctance of the majority of the Procedures Committee to approve the proposal mentioned in the Appendix to this paper is another illustration. People and their representatives generally want two incompatible things: more state services (which cost money) and lower taxes. Greater efficiency in the use of resources will not reconcile these objectives. Up to a point people accept this and, so long as they do, control over expenditure in a democracy will be weak. Only when they call a halt, as in the Proposition 13 case, will expenditure control become more effective, but in a world of rising living standards this is likely to be a temporary phenomenon. Hard times are a good expenditure control but who wants them? In good times why should not the state do that little job we want it to do? In modern states most of us pay a greater proportion of our incomes in taxes then our grandfathers did but what we have left still enables us to live more prosperously than they did, as well as having more state services available to us than they had available to them. We have our cake and eat it because, of course, our economies have advanced as a whole and, so long as that phenomenon lasts, the advocate of stricter public control of expenditure is likely to lead a minority. He, not the advocate of more spending, seems to be the impractical idealist. The realist, who gets the votes, is the man who wants more spending on more services, giving more jobs generating more taxes enabling more spending, and so on, but to discuss that argument would go far beyond the bounds of this paper. We must remember that one's view of it will colour one's view of public expenditure control and so far, in most developed countries, the spenders are winning.

APPENDIX

MINORITY PROPOSAL OF THE PROCEDURE COMMITTEE

Each committee would consider the Estimates referred to it, and would be empowered to recommend to the House, within a specified period, either that the Estimates be approved or that the Estimates be approved with such changes as the committee consider desirable. If a committee failed to report within the specified period, it would be deemed to have reported that the Estimates be approved. If, however, a committee took advantage of the opportunity to report recommendations concerning the main Estimates referred to it, the following procedures would apply:

(i) If it recommends that any of the Estimates referred to it should be increased such recommendations will be regarded as of an advisory character only and the Committee shall be deemed to have recommended the approval of the Estimates concerned. This provision will avoid any conflict with the established rules relating to proposals for increases in public expenditures.

(ii) If it recommends, or is deemed to have recommended, the approval of the Estimates referred to it, the Question on any Motion in the House to approve those Estimates will be put forthwith without amendment (unless proposed by a minister of the Crown) and without debate.

(iii) If it recommends a reduction in any of the Estimates referred to it,

(a) if the government accepts the recommendation, any Motion to approve the Estimates, so amended, will be put forthwith, without amendment or debate; but

(b) if the government opposes the recommendation, any Motion to approve the Estimate concerned will be debatable, but any Question necessary to dispose of proceedings thereon will be put at midnight or two hours after they have been entered upon, whichever is the later;

(iv) In order to avoid an unlimited number of debates of the kind specified above, all the Estimates referred to each Committee will be embodied in a single Motion, ensuring that, at the most, only one debate (limited to two hours, unless the government or opposition find time before 10:00 p.m.) will take place on the recommendations of each committee.

A similar procedure, subject to a much shorter timetable, could be introduced in respect of supplementary Estimates. The above procedure would, of course, not be in derogation of the present right of the opposition or of other members to move reductions in the Estimates.

APPENDIX

MINORITY PROPOSALS OF THE PROCEDURE COMMITTEE

Chapter Seven

The Public Evaluation of Public Spending: The American Experience

by
Dean Crowther

The American experience with the control of public expenditures and the use of program evaluation is probably not the best model in light of the occurrences of the past decade including Watergate, increasing inflation, high unemployment, rapidly increasing government expenditures, and high deficits and pressures by citizen-taxpayers with regard to high taxes. On the other hand, perhaps this experience allowed us the opportunity to learn very positive lessons. The pressing need for more accountability in government, and the strong need for political fiscal responsibility in holding down spending and, in fact, reducing taxes in response to a taxpayers' revolt, are two important concerns.

I will first cover some brief historical scene setting and describe the events leading up to some rather major changes in American expenditure controls, including the *Congressional Budget Act*, consideration of sunset legislation requiring program evaluation, and related matters, such as zero-base budgeting.

Public spending with evaluation and accountability of funds are the essence of a full society. The U.S. budget is presently in the $500 billion range and the deficit is in the $50 billion range. These large numbers not only affect the U.S. economy but they also have some effect on the world economy. Measurement and tracking of this impact is of critical importance.

Shortly after the U.S. Constitution was drafted in Philadelphia in the late 1700s, Benjamin Franklin was asked what form of government he and the other "founding fathers" had designed. Franklin replied: "Republic, if you can keep it." Franklin's caveat is pertinent today, particularly in our consideration of the topic "The Public Evaluation of Public Spending." The American republic rests on a central assumption made by the "founding fathers" nearly two hundred years ago. The assumption was that citizens, through their elected representatives, can guide and control their government. For the republic to be kept, citizens, through their representatives, must retain this ability. With all the changes that have since come about, it is clear that to guide government and make rational decisions, the public, the Congress, and the president will need evaluation of the actual performance of government programs.

Before I go much further, let me be sure to mention the close sharing of technical methodologies and approaches between the U.S. General Accounting Office and both the Auditor General's Office and the new Comptroller General's Office of Canada. Representatives of both organizations have tried, and I believe successfully, to learn from each other's experiences in terms of managing and assessing public programs. These linkages are, of course, very important and we are dedicated to assist where we can to make them even stronger.

FRAMEWORK OF THE U.S. POLITICAL SYSTEM

The structure of the U.S. government is designed to assure representation of various constituencies and to prevent the aggrandizement of power by one individual or faction. It provides for a limited federal government, the vertical separation of powers among the legislative, executive, and judicial branches, and various terms of office and methods for the selection of individuals to hold office in each branch. The U.S. Constitution also provides for the horizontal separation of power between the federal and state governments. In addition, the first 10 Amendments to the Constitution, the "Bill of Rights" further limits the power of government by reserving certain rights to individual citizens.

The general framework of the U.S. political system provides the American public a means of expressing their opinion and, by casting their vote, offer their evaluation of the government's performance. Recently, this was exhibited in the State of California in the form of a taxpayers' revolt called "Proposition 13" in which the voters loudly and clearly voted to sharply reduce property taxes in California, thereby reducing state government spending. Also, information is provided to the public, including government officials, through the operation of a free press. The information provided by the press, coupled with information received from other sources, gives the public some basis, although rather limited, for making its evaluations of the end results of the performance of their representatives and their government. Although the press and related information are helpful to the public to decide upon how well government is working, such information is not sufficient for government managers to use in making operating and policy decisions. Accordingly, new and better sources of information and new methods for evaluation have had to be developed to supplement these basic mechanisms of public evaluation

CONGRESSIONAL INVESTIGATION: A SURROGATE FOR PUBLIC EVALUATIONS

From the outset of the American republic, congressional investigations have served as a surrogate for public evaluations of government performance. The conditions favouring the use of this method of "public" evaluation include the growth of administrative government, the inability of the press to monitor the internal actions of government administration, and the willingness of the Congress to exercise the power of legislative inquiry implied by the Constitution.

The first congressional investigation was begun in 1792 when a House select committee examined the defeat of American troops on the Ohio frontier by Indian tribes. Such investigations have been conducted ever since, and have been increasingly undertaken since the turn of this century.

Congressional investigations have also been undertaken in connection with the exercise of the power of appropriation. The diary of James A. Garfield, chairman of the House Appropriations Committee between 1871 and 1875, and later president of the United States, is indicative of how the appropriations committee viewed their role just over one hundred years ago. Mr. Garfield states that the Committee on Appropriation looks inward upon the machinery of the government and reviews in detail all its various functions. Congressional investigations are now carried out by a number of committees in both Houses as well as being augmented by program evaluation and policy analysis by the General Accounting Office, and specific analysis and studies performed by the Congressional Research Service (CRS), Congressional Budget Office (CBO), and the Office of Technology Assistance (OTA).

THE GROWTH OF BUREAUCRACY AND BUDGET

Initially, only four executive departments (War, State, Navy and Treasury) were created by the Congress in the eighteenth century. It was not until 1849 that the Department of Interior was added. Then, near the end of the nineteenth century, the growth of the bureaucracy began to accelerate significantly. This growth has continued, particularly during the New Deal, with the creation of a host of agencies to deal with the effects of the economic depression of the 1930s. Today, there are a number of federal departments that employ in excess of 100,000 persons and two departments—Defense and Health, Education and Welfare—that are responsible for expenditures in excess of $100 billion per year each.

The massive growth in the administrative functions and the budget of the American government was not foreseen by the founders. This has led to political conflicts over the location of administrative agencies in the U.S. system and the line of accountability applicable to such "independent" agencies, especially for budget purposes.

The Congress retains the power to create, terminate, design, fund, and oversee such agencies. The president is responsible for the operation of these agencies. The executive branch exercises varying degrees of control over the agencies, depending on the degree of independence provided to them by the legislation (e.g., Federal Reserve, other regulatory and quasi-judicial agencies, and a number of off-budget agencies which receive little control by the executive branch).

EVALUATION METHODS

With the growth of administrative bureaucracy in the United States, new evaluation methods had to be developed. The *Budget and Accounting Act* of 1921 set up the president's budget system, created the old Bureau of the Budget (which performed some evaluation functions for the president) and created the General Accounting Office (which performs evaluation functions for the Congress). Essentially, the *Budget and Accounting Act* of 1921 was a statement of the Congress that better discipline was needed in the budget process, but the Congress itself was not able to exercise this discipline; and the audit function should be located in the legislative branch so that it could be independent of the department or agency under audit.

The growing size and scope of government requires that agencies establish internal audit and evaluation organizations to monitor themselves, since monitoring of the parts is an integral component of self-steering. Otherwise, GAO would have to become a massive monitor. Various amendments to the 1921 *Budget and Accounting Act* have served to emphasize the need for each executive branch agency to develop its own internal audit and management control systems. GAO has attempted to help agencies meet this objective.

Government Accounting Office

This is a good place to pause and divert for a moment and explain briefly the organization and function of GAO and the other arms of the legislative branch. The U.S. General Accounting Office is headed by the Comptroller General, Mr. Elmer B. Staats, who serves for a fifteen-year term. This Office is responsible for reviewing government activities and preparing reports to the Congress reporting on the effectiveness of operating programs and recommending improvements that can be made. The Office has professional staff located in most government agencies in Washington where they perform their reviews, staff in fifteen regional offices throughout the United States, and in selected locations abroad. The staff responds to congressional requests to perform specific reviews and performs selected comprehensive program evaluations at its own initiative. Also under the *1974 Budget Act*, GAO assists the Congress in its congressional budget activities. GAO has a total staff of about 5,200.

Congressional Budget Office

The Congressional Budget Office was established by the *1974 Budget Act* and is instrumental in the Act's operation in Congress. The CBO works primarily with the House and Senate Budget Committees, but also performs policy analysis for any congressional committee upon request. The CBO provides the staff assistance required to develop the budget resolutions which are voted on by both Houses of Congress and sets the spending limits, total budget authority, and the amount of the deficit.

Additionally, the CBO keeps score of the budget activities of each Committee during budget deliberations so the budget committees can be certain the budget stays within the approved ceilings. The CBO does economic analysis, revenue estimating, tax expenditure analysis, and specific policy analysis. The CBO has a staff of about 200.

Congressional Research Service

The Congressional Research Service is the research and analysis arm of the Library of Congress. The CRS responds to congressional requests from all committees and members. It is responsible for a large number of studies that provide direct input to committee deliberations on almost any subject. It also provides direct staff assistance to committees upon request. The CRS has a staff of about 900 and includes a wide range of disciplines and backgrounds. There are also a number of senior specialists in selected fields that provide a high level of professional expertise on given subjects and program areas.

Office of Technology Assessment

The Office of Technology Assessment was established to provide research, analysis, and evaluation of selected science and technology areas. It has a staff of about 40 persons and contracts for a number of studies to be performed in the science and technology areas. Its purpose is to provide studies in the science area to the committees involved in such matters.

In the early 1970s, the Congress was stirred to action to reassess its own role in the federal budget priority setting by (1) Watergate and its impact calling for better accountability; (2) a series of major presidential impoundments of funds for programs the Congress especially wanted carried out; (3) the realization that a large portion of the budget was not subject to appropriation action and control because of large, continuing, open-ended entitlements and related types of programs; and (4) the powerful act of frequently raising the debt ceiling without being able to relate the increase to any particular congressional revenue or expenditure decisions.

Recognizing these diverse problems, Congress considered a wide range of alternative reform approaches and in July 1974, enacted the *Congressional Budget and Impoundment Control Act* of 1974 (Public Law 93-344). This Act popularly referred to as the *1974 Budget Act*, established budget committees in each House, and a Congressional Budget Office to administer the new budget processes, and established a set of procedures and a timetable for each year's budget activities.

The *Congressional Budget Act* of 1974 provides a mechanism for the first time with which the Congress can set spending levels for the federal budget as a whole. Under that new law, congressional attention is now focused on total spending levels for the entire government, as well as on broad categories for budget functions which, taken together, constitute a comprehensive but general statement of national priorities.

Early in each session, all of the committees review the president's budget proposals for the next fiscal year and report their views to the House or Senate Budget Committee. The budget committees, considering the overall economic condition of the nation and the views of the committees and the president, recommend to the House and the Senate overall revenue, spending, and debt targets, and targets for each of the budget functions, such as national defence, energy, health, and general government. These recommended levels are acted upon by the House and the Senate through the normal legislative conference process which is referred to as a concurrent resolution.

As Congress acts upon spending bills for individual departments and agencies, the Congressional Budget Office keeps score against these targets. Any bill that would take Congress over one of its targets is subject to special parliamentary procedures; thus, Congress can go over a target but it can only do so knowingly. In September, the Congress reviews the spending actions it has taken and sets firm ceilings on the budget totals and each of the budget functions in a second concurrent resolution which, together with the individual departments and agencies spending acts, comprise the federal budget for the next fiscal year.

This year, for example, the president's budget for fiscal 1979 proposed a spending level of $500.2 billion with a revenue estimate of $439.6 billion which would result in a budget deficit estimate of $60.6 billion. After completion of the second concurrent resolution on the budget in September, the Congress reduced the President's budget estimate of $500 billion to $487.5 billion, changed the revenue estimates as a result of a difference in the amount of a proposed tax cut from a total of $439.6 billion to $448.7 billion and the resulting deficit estimates from $60 billion to $38.8 billion. Those are significant changes that probably would not have come about without the pressures brought to bear on the Congress by virtue of the *1974 Budget Act*. Of course, some of these numbers will change now as a result of recent (October 1978) tax reform legislation that resulted in an $18 (plus) billion tax cut.

Today, the Congress is in its third year of full operation under the budget act process and has been somewhat successful in adhering to the discipline of the process. There has been a strong commitment by the congressional leadership, the budget committee members, and the other committee members to "making the process work" and for the most part it has.

The major confrontations and strains on the process have been over very tough national policy issues that would heavily strain any process: defence versus social program spending, federal financing of abortions, and federal support for agriculture. To the extent the process does focus the Congress' attention on the major policy issues and their budgetary impacts, the process is obviously working. From what has been seen in the budget debates and actions, there is much more attention to the budgetary implications of policy decisions throughout the legislative process and a great deal more understanding of, and interest in, the budget process by all members of the Congress, the news media, and the public as a result of the *Budget Act*. The Act reflects the U.S. Congress' awareness of the need to consider the budget in conjunction with the overall U.S. economy, the importance of considering the effects the various budget levels will have on macro-economic policy as a tool for meeting U.S. economic objectives (e.g., controlling inflation, increasing aggregate demand), as well as the critical need to discipline its own spending through the establishment of budget targets as has been done to some degree in the president's budget process since 1921.

The institution of zero-base budgeting in the president's budget process within the U.S. Executive Branch, on the other hand, reflects a new emphasis on the "bottoms up" approach to improving budget decisions—each of the "pieces" are individually justified and ranked as an aid to determining the appropriate level of funding for programs. Zero-base budgeting in the executive branch is now in its second year and the results, or success of the process, are still subject to evaluation.

OTHER STEPS TO IMPROVE EVALUATION

In the 1970s the Congress has taken other steps to improve its capability to oversee the federal government. Besides implementing new procedures for budget making, the Congress has increased its staff support for members and committees, created two new committees (the House and Senate Budget Committees), and two new congressional support agencies (the Congressional Budget Office and the Office of Technology Assessment), and expanded the analytical functions of the Congressional Research Service and the General Accounting Office. In many pieces of legislation in the 1970s, the Congress has included requirements for programs to be periodically reauthorized and for agencies to study and report to the Congress on the performance of programs authorized by the legislation.

During the 1970s, the Congress has also expanded the oversight responsibilities and authorities of its standing committees. In 1970, the committees were required to report on their oversight activities during each Congress. Each House has also adopted new rules giving committees additional oversight responsibilities.

Various legislative proposals for further improving oversight and accountability of federal programs are currently under consideration in the Congress. For example, one proposal just enacted into law on October 14, 1978, provides authority to establish an office of Inspector and Auditor General in each of the major federal departments. Another proposal would require the president to make annual reports on the performance of each federal program, including an assessment as to whether or not the programs are working. Perhaps the most comprehensive oversight reform proposal under consideration is Senate Bill (S.2), the *Program Reauthorization and Evaluation Act* of 1978, popularly referred to as ''Sunset'' because it requires most programs to terminate every ten years unless completely reauthorized. The July 13, 1978, report on S.2 by the Senate Committee on Rules and Administration (Senate Report 95-981) states that:

> The purpose of the bill, as reported by the Committee on Rules and Administration, is to improve the effectiveness and efficiency of the Federal Government by strengthening congressional procedures for the review and reauthorization of Federal programs.

This bill would require substantial periodic evaluation of all government programs before being considered for reauthorization.

PROSPECTS FOR THE FUTURE

Given that the economies of the world will likely continue to become more interdependent upon each other, it seems reasonable to expect the continued development of more sophisticated management control mechanisms and aid for government managers. Also, I would expect more pressing demands from Congress to monitor and control program costs in line with program performance, given that inflation will continue to eat away at the ''true'' buying power of U.S. dollars, and continued pressure will be brought to bear by taxpayers and voters to reduce taxes and reduce government expenditures.

I also would expect continued pressure to be exercised by U.S. government managers to keep the size, scope, and cost of government lean and trim. We all know there is growing concern that our ''self-steering'' ability has been reduced because of increases in the size, scope, and complexity of the U.S. federal government, the perceived uncontrollable nature of the U.S. federal budget, and the diminishing portion of the U.S. federal budget available for new programs.

The Congress, the president, and the public are increasingly removed from the ''helm'' of the government in that their ability to control all of government seems to have been diminished over time because of several factors: (1) the vast number of federal programs; (2) the dramatic increase in the percentage of federal spending for so-called uncontrollable programs; (3) the even more rapid growth in the costs of federal programs' permanent appropriations; (4) the complexity of the social, economic, ecological, technological, and international conditions government is attempting to influence; and (5) the complexity of the government's operations and programs themselves. The combined effect of these factors is the growing impression by the citizens of the difficulty of managing such large and complex programs as if the ship of state is almost always in troubled waters. I believe this impression, unless appropriately countered, can result in a decline in public confidence.

These factors require additional insight and ingenuity in program management and they require considerably more evaluation to be performed. I believe Congress will be called upon to make some of the most difficult program and budget decisions ever in order to meet the ever-changing national priorities. These decisions will certainly require better evaluations of how programs are working and proposals for making them work better. These factors have spawned additional oversight evaluation reform proposals, including "sunset" tax reform, and many social program reforms.

To summarize, as a result of a number of factors, probably beginning with the energy crisis and Watergate, major reforms, including governmental evaluation and budget structures, and others being developed in the United States, will result in requirements for better oversight, evaluation, and control of federal programs and control of federal spending which are all critical to the health and well-being of the U.S. economy. Such reforms to better control government spending and to better utilize resources will not be brought about without considerable turmoil and agony, but their necessity far outweighs the chaos that can result without such controls.

Chapter Eight

Bureaucratic Growth in Canada: Myths and Realities

by
Richard M. Bird

and
*David K. Foot**

Few questions may seem simpler to answer than the following: "How many people work for the government?" In fact, however, it turns out to be far from easy to answer this question, both because it is unclear exactly what those who ask it mean by 'government' and because different sources give different numbers for what would seem to be the same concept. Matters become even more confused when one tries to find out how fast government employment has grown or just who all these people who work for government are and what they do. Almost two decades ago, the Glassco Report complained that the data available on public sector employment in Canada were confusing and inconsistent (Royal Commission . . . 1962, p. 350). Matters have not changed for the better since then, as any regular newspaper reader may have noticed. Nevertheless, available data do permit one to obtain a much more coherent and complete picture of the size, growth, and composition of public employment than seems generally to be realized. This paper draws on these data—most of which are reported in much more detail in Foot (1978)—to show that the reality of bureaucratic growth in Canada is in many important respects different from the myths perpetuated in much popular commentary. After a brief note on the different concepts of the public sector used here, the paper discusses in turn what the data show about the size, growth, and composition of public sector employment in Canada. A brief concluding section contrasts the 'reality' shown by the numbers with the 'myths' that seem to dominate popular discussion of this subject.

---

* This paper draws heavily on work done as part of a much broader project on public employment carried out at the Institute for Policy Analysis for the Institute for Research on Public Policy. We are grateful to our colleagues on this project for their contributions to the work as a whole, but we are ourselves solely responsible for what is said here. Earlier versions of some of the material in this paper were circulated as "Trends in Public Employment and Pay," Policy Report No. 1 (Montreal: Institute for Research on Public Policy, November 1977) and presented at the Thirty-Fourth Canadian Economic Policy Committee Meeting of the C.D. Howe Research Institute, May 1978.

CONCEPTS OF THE PUBLIC SECTOR

Discussion of the subject of public employment is bedevilled by the propensity of discussants to use different numbers, usually from different sources, to refer to what is allegedly the same phenomenon. This paper makes use of data from four principal sources: the census, the *Taxation Statistics* issued annually by Revenue Canada, a series of publications on government employment issued by Statistics Canada, and information from the federal and provincial civil service commissions. Each of these sources uses different concepts and definitions—see the detailed discussion in various chapters of Foot (1978)—and the figures from one source are not easily comparable to those from another. Moreover, the various definitions have often changed over time, a fact which makes the calculation and comparison of growth rates particularly tricky. Finally, there are still other numbers on 'public employment' floating around—from labour force statistics, the federal *Estimates*, and so on—thus confusing matters even more (Bird 1978).

This paper (except for the Appendix) spares the reader a detailed statistical catalogue of the many adjustments that have had to be made to the data used here. Nevertheless, in view of the inherent complexity of the subject matter, it is not always easy to keep straight exactly what is being discussed at each point. In the hope of helping guide the reader through this maze, a number of terms that will be used as consistently as possible throughout the paper are therefore introduced.

The narrowest concept of public employment used here is that of *civil service employment*. 'Civil servants' are defined as those who have been "appointed to a position on a full-time basis and whose entry into government service has been subject to final certification by a central personnel agency" (Hodgetts and Dwivedi 1974, p. 180). Thus defined, it is clear that this term applies only to what may be thought of as the 'hard core' of the federal and provincial bureaucracies. When some non-civil service but regular federal and provincial employees, together with the employees of municipal governments, are added to civil service employment, the resulting total is called *government employment*. Finally, there are a large number of people employed by the state on a regular basis who would not be included in government employment thus defined. Examples are teachers and other school board employees, hospital workers, and the employees of government enterprises. The broadest concept of direct public sector employment used here, which encompasses such groups, is called *public employment*. (Sometimes the term 'public sector employment' is used when it does not really matter what specific concept is being referred to.)

These concepts may seem simple enough, and so they are, in principle. Life is seldom really simple, however; certainly not the life of anyone who tries to work with Canadian public sector data. There are, for example, other concepts of public sector employment, such as that used in the census, that do not readily fall into the above classification. More importantly, for some purposes it may make sense to include, say, the armed forces in the concept of 'government employment' or enterprise employees in the concept of 'public employment' while for others it may not. Data sources also differ in the extent to which they include different groups in the public sector; Statistics Canada, for example, includes enterprise employees while the taxation data do not. Furthermore, it is not always clear in the sources whether the numbers refer to full-time employees or full-time 'equivalents', whether regular part-time employees or casual employees are included, and so on. For some purposes it may matter a great deal whether a 'teacher' for example, is a real, living human being or a statistical abstract made up of 'x' days each worked by, let us say, ten supply teachers. Although every effort is made to refer as consistently as possible to the three major concepts defined above and to signal significant deviations, readers must therefore stay alert and constantly ask themselves "now what, exactly, does this concept *mean*, and is it the one I am really interested in?" (Those who want more details, or to rearrange the numbers their own way, will find the basic data in Foot (1978) for the most part.)

THE SIZE OF PUBLIC SECTOR EMPLOYMENT

Table 1 contains an estimate of public sector employment in 1975, for the three concepts defined in the previous section. Several interesting conclusions may be immediately deduced from these numbers.[1] In the first place, less than one quarter of all public employees are 'civil servants', with all that is commonly taken to imply for job security, methods of selection and promotion, and so on. Secondly, while more than half of all civil servants in Canada work for the federal government, that government is in total by far the *least* important employer of the three levels of government. The provincial level (including hospitals and post-secondary education) is the most important by a considerable margin, and the provincial and local levels together account for three quarters of all public employees in this country. Finally, the 'typical' public employee—if such a concept has any meaning—is more likely to work for a hospital or educational institution than to be a direct employee of *any* of the three levels of government (excluding government enterprises).

The total shown in Table 1 is quite close to that which may be derived for public employment from the quite different basis of tax returns (Foot and Thadaney 1978). Indeed, it may even be argued that the slightly larger *Taxation Statistics* figure of 2.4 million 'public employees' in 1975 is more

---

[1] Actually, some of these conclusions really draw on the rearrangements of the basic data shown in Appendix Table A2. The figures in Table 1 differ slightly from those in Bird (1978), mainly because the latter reported mainly year-end figures—and there is a significant seasonal pattern in government employment (Foot, Scicluna and Thadaney 1978a).

Table 1
PUBLIC SECTOR EMPLOYMENT IN CANADA, 1975
(Thousands)

1. Federal civil service	273.2	
2. Provincial civil service	259.9	
(Total civil service employment)		(533.1)
3. Other federal civilian	41.1	
4. Armed forces	79.8	
5. Other provincial	115.1	
6. Municipal	251.3	
(Total government employment)		(1,020.4)
7. Education	525.1	
8. Hospitals	369.5	
9. Enterprises	328.1	
Total public employment		2,243.1

Sources and Notes: see Appendix.

representative of the actual scope of public sector employment than the figures in Table 1, on the grounds, for example, that most non-profit institutional and educational organizations—including those excluded in Table 1—likely derive the bulk of their support from the public sector. Furthermore, the taxation figures may well capture more completely various casual and part-time public sector employees who are included only in part in the figures in Table 1. On the other hand, the taxation data exclude employees of government enterprises. There is no way of knowing the extent to which these various inclusions and exclusions offset one another, and on the whole the data in Table 1 seem to provide the most reliable guide to the size of public sector employment.

This estimate of the size of public employment in Canada is, on balance, probably rather conservative. In particular, if one is prepared to extend the definition of 'public sector employees' to include all those who derive their income wholly or partly from work for government, one would also have to include as 'indirect' government employees many persons employed by, say, aerospace firms and paving contractors who work for governments, since their jobs are basically generated by public purchases of goods and services from the private sector. A crude estimate—based on simple extrapolation of the work reported in Bucovetsky (1979a)—is that perhaps 685,000 people were thus 'indirectly' employed by government in 1975. The total employment directly and indirectly attributable to the public sector (including a very crude estimate of 150,000 casual employees not included in Table

1[2]) in 1975 may be estimated, very roughly, at around 3.1 million persons. A really enthusiastic counter might then move on to include at least another 2.9 million adults (recipients in 1975 of Unemployment Insurance and Old Age Pensions[3]), thus arriving at a grand total of six million adults dependent on public sector activity for their primary financial well-being.

Although this last point takes us well beyond the limits of the present discussion, it is, to say the least, clear that the public sector is a *very* large employer in Canada. Indeed, it is far from clear that even those who must decry the size and growth of government really understand how important it is in this respect. The major conclusion emerging from this section, then, is that at least 2.2 million people—24 per cent of the employed labour force—worked *directly* for the public sector in 1975, the largest proportion of them in social services (health, education) and at the provincial and local levels. It is thus a reality that the public sector is a very important employer in Canada, and it is also a reality that most public employees work in the 'social' services. It is, however, a myth that most public employees work for the federal government, and it is also a myth that most public employees are civil servants.

THE GROWTH OF PUBLIC EMPLOYMENT

Census Data

Although the concept of employment in public administration used in the decennial census is considerably narrower than any of those mentioned to this point (except civil service employment), this source provides the only available long-term data on public sector employment and therefore affords a valuable perspective on more recent—and better-documented—developments. From 1911 to 1971, public sector employment as recorded in the decennial census grew steadily (apart from the wartime bulge in the armed forces) at an average annual rate of over 3 per cent, increasing from 7 per cent of total (census) employment in 1911 to 12 per cent in 1971 (Bird 1978). Public employment grew most rapidly, however, during the 1940s and 1950s. From 8 per cent of employment in 1931, public employment rose to 10 per cent of the total in 1951 (after a wartime peak of 11 per cent) and 14 per cent in 1961. Although public employment continued to grow during the decade from 1961 to 1971, employment in the private sector expanded so rapidly in this period that public employment as a percentage of total employment actually *fell* to 12 per cent in 1971. The fastest-growing component of census public employment in the post-war period was

---

[2] See Bird (1978), note 14

[3] Statistics Canada, *Social Security: National Programs* (86-201). Not all these people received these payments all year or were heavily dependent upon them for income: but there are enough other transfer programs to make it not unreasonable to use those figures as a proxy for total 'state dependents'.

provincial government employment, which more than doubled over the period in relation to total employment.[4]

The census data thus suggest several interesting conclusions. First, the fastest growth in public employment (relative to total employment) took place in the early post-war period. Secondly, only provincial government employment continued to grow rapidly (in relative terms) in the 1960s. Finally, the public sector as a whole was *less* important as an employer in 1971 than it had been a decade earlier. It is thus a reality that the public sector is a much more important employer now than it was 50 or 60 years ago, but it is a myth that public employment has grown particularly rapidly in recent years, and it also appears to be a myth that federal government employment has in any sense led the growth of public employment.

Taxation Data

A more comprehensive source of time-series data on public employment in Canada is the annual *Taxation Statistics* published by Revenue Canada, Taxation. Over the post-war period as a whole, total 'public' employment as recorded in this source grew at an average annual rate of close to 7 per cent, rising from approximately 13 to 20 per cent of total individual tax returns filed (Foot and Thadaney 1978). Once again, however, the fastest relative growth appears to have taken place in the period before 1961. Although the absolute number of returns filed by public employees continues to grow at a rate of over 6 per cent a year thereafter, the relative size of public employment peaked at around 21 per cent in 1971, and has actually fallen slightly in more recent years. These trends hold, roughly, for both the 'government employment' and 'public employment' concepts as measured by these data.

The taxation data thus support the earlier inference from the census data that it is a myth that the last decade has been a period of especially rapid growth in public sector employment—and also the suggestion that the reality in the 1970s may have been a cessation in the relative growth of public sector employment as a whole. These data also support the conclusion from census data that federal employment has grown more slowly in recent decades than provincial employment. Over the post-war period as a whole, the federal government, initially the largest employer in the taxation data, has declined sharply in relative importance, while the 'institutional' sector (which includes, for example, hospital employees) has grown so rapidly that it has become by far the largest 'public sector' employer. The provincial and municipal sectors, especially the former, similarly grew in relative terms

---

[4] To some extent, of course, this rapid expansion of provincial activities was presumably facilitated by the growth in federal transfers to the provinces, especially in the 1960s.

over the period, while the 'educational' sector declined slightly in importance.

Civil Service Data

Neither census nor taxation data thus lend any support to the popular perception of rapidly expanding public sector employment, particularly at the federal level, in recent years. Another source of long-term data on government employment is provided by the annual reports of the federal and provincial civil service commissions. Although these data are not readily comparable either among jurisdictions or over time owing to (sometimes marked) differences and changes in coverage, they are worth special note here both because 'civil servants' are, so to speak, the hard core of public employment and because if one asks the various governments "how many civil servants do you employ?" these are the numbers they generally produce.

Federal data, which are available back to 1912, show that the average annual growth of the federal civil service was considerably higher in the period before 1951 than in the last two decades, thus lending some general support to the earlier argument from census data about the time pattern of public employment growth (Bird 1978). From 1912 to 1951, for example, federal civil service employment grew at an average annual rate of 4.6 per cent, compared to only 3.0 per cent in the subsequent 25 years. In contrast to the trends discussed earlier, however, these data suggest that the 1950s were a slow growth period for the federal civil service, while the rate of growth increased considerably in the 1960s and, especially, in the early part of the 1970s. In the 1971 to 1976 period, for example, the federal civil service expanded at a rate of 5.2 per cent, or faster than over any earlier comparable period.[5]

Although the provincial civil service data are in a less satisfactory state, for the most part they show lower rates of growth in the 1960s compared to the 1950s (though higher than the federal rate in both decades), with no marked trend in the early 1970s. In total, there has clearly been a much greater increase over the last two decades in the size of the provincial than of the federal civil service, despite the sharp increase in the recorded number of federal civil servants in 1967 (when a new *Public Service Employment Act* was passed). Total civil service employment in Canada rose steadily during the post-war period—from approximately 3 per cent of the labour force in 1951 to 4 per cent in 1961 and 5 per cent in 1971, remaining at about that level thereafter.

---

[5] The taxation data too show an increase in the rate of growth of federal employment (not civil service) in the early 1970s, although it is still below that for other levels of government (Foot and Thadaney 1978).

Although the federal and provincial civil services together constitute less than one quarter of total public employment (as defined in Table 1), these data are particularly interesting since they appear to conform much more closely to the apparent public perception of what has been going on than do any of the broader aggregates discussed earlier. Indeed, the *only* segment of public employment that has clearly expanded at a faster rate than 'normal' in the last few years appears to be the federal civil service, thus confirming the casual impression that many Canadians tend to judge the performance of public sector employment as a whole by focusing on this one small, and to a considerable extent atypical, segment of it.

Moreover, while federal employment as a whole may indeed be relatively less important now than a few years ago, as argued above, the proportion of federal employees subject to civil service rules is now higher than ever. In 1975, for example, 87 per cent of federal civilian employees were 'civil servants', with all that is usually taken to imply about job security, pensions, and so on, compared to only 58 per cent in 1961. At the provincial level too—which of course remains much the more important in terms of total employees—the proportion of government employees included in the civil service appears to have grown in recent years. Since it is these figures which tend to receive most publicity in the media, this phenomenon may have strengthened the apparent general public perception of the growth of public sector employment as a whole. In other words, the reality is that the federal civil service did grow quickly over the last decade and that an increasing number of government employees at both federal and provincial levels are civil servants. It is, however, still a myth that these trends represent fairly what has been going on in the public sector as a whole.

Statistics Canada Data

Finally, although the Statistics Canada data presented in Table 1 are available even on an estimated basis only since 1961, they clearly confirm the impression that the rise in public employment has been amazingly small in recent years in relative terms. The detailed estimates are presented in the Appendix to this paper and summed up in Table 2. Although the public sector, as a whole, employed almost 900,000 more people in 1975 than in 1961, total employment in the country increased by 3.3 million over the same period, with the result that the share of public employment accounted for by the public sector rose only slightly, from 22.2 to 23.7 per cent.

Public employment is indeed very important in Canada, as noted in the previous section; but all the available data suggest that it is a myth that its relative importance has changed much in recent decades—with the notable exception of civil service employment. Moreover, throughout the period covered by the comprehensive data reported in the Appendix (1961 to 1975), the provincial level of government (broadly defined to include hospitals and post-secondary education) has been by far the largest employer and has accounted for almost all of the growth. The municipal (including schools) share of the total has remained almost constant at around 31 per cent, while the federal share has declined sharply from 34 to 24 per cent. It is thus also a myth that the federal sector has become relatively more important as an employer in recent years.

PUBLIC EMPLOYMENT IN PERSPECTIVE

The absolute numbers of public employees recorded in Table 1 may, or may not, appear impressive. In general, they should *not* appear very impressive unless they are compared with something else which puts them into perspective. Comparison with data on total employment, as in Table 2, is one obvious way to do this. Another appropriate comparison, since most public employment is clearly in the service sector, is with service employment. Table 3 makes such a comparison.

Table 2
THE SHIFTING COMPOSITION OF PUBLIC EMPLOYMENT, 1961-1975
(As per cent of total employment)

Year	Federal Government	Provincial Government	Municipal Government	Education	Hospitals	Enterprises	Total
1961	5.2	2.7	2.3	4.7	3.4	3.8	22.2
1962	5.2	2.7	2.3	4.9	3.5	3.7	22.2
1963	5.0	2.8	2.3	5.0	3.7	3.6	22.3
1964	4.8	2.8	2.3	5.1	3.8	3.5	22.3
1965	4.6	2.8	2.3	5.1	3.9	3.5	22.2
1966	4.5	2.9	2.3	5.3	4.0	3.5	22.4
1967	4.5	2.9	2.3	5.5	4.2	3.6	23.0
1968	4.4	3.1	2.4	5.7	4.2	3.4	23.2
1969	4.3	3.1	2.4	5.9	4.2	3.4	23.3
1970	4.2	3.2	2.6	6.1	4.2	3.3	23.6
1971	4.3	3.4	2.5	5.8	4.2	3.3	23.4
1972	4.3	3.4	2.6	5.7	4.1	3.3	23.3
1973	4.2	3.8	2.6	5.6	3.8	3.3	23.2
1974	4.2	3.9	2.6	5.5	3.8	3.4	23.4
1975	4.2	4.0	2.7	5.5	3.9	3.5	23.7

Note: Rows may not add to totals because of rounding.
Source: Appendix.

The most interesting fact that emerges from this table is that, although public employment has increased slightly as a proportion of total employment (Table 2), it has actually *declined* as a proportion of service sector employment. This reflects the fact that service sector employment increased more as a share of total employment—from 54 to 64 per cent over the period covered in Table 3—than did public employment. The rapid increase in the importance of tertiary activities in the Canadian economy is, of course, well known: what seems to be much less well appreciated, however, is the fact that public employment, broadly defined, has accounted for less and less of this rising total. Over the 15 years from 1961 to 1975 four out of every five new jobs created in Canada were in the service sector—but only a little more than one in four was in the public sector, broadly defined, less than one in eight in the government sector (as defined earlier), and only two in every hundred in the federal government. It is thus a myth to attribute the dawning of the 'service economy' (Fuchs 1968) to the growth of the welfare state.[6]

Table 3
PUBLIC EMPLOYMENT AS A PER CENT OF SERVICE EMPLOYMENT, 1961-1975

Year	Federal	Provincial	Municipal	Education	Hospitals	Civil Service	Total Non-Enterprise Civilian	Total Public Employment[a]
1961	6.1	5.1	4.3	8.9	6.4	7.2	31.0	40.4
1962	6.0	5.0	4.3	9.2	6.5	7.2	31.0	40.2
1963	5.6	5.0	4.2	9.0	6.7	7.0	30.6	39.2
1964	5.4	5.0	4.1	9.2	7.0	7.1	30.7	39.1
1965	5.3	5.0	4.1	9.1	7.0	7.0	30.4	38.4
1966	5.3	5.0	4.0	0.2	7.0	7.3	30.5	38.1
1967	5.3	5.0	4.0	9.3	7.1	8.5	30.8	38.4
1968	5.1	5.2	4.0	9.6	7.1	8.5	31.0	38.2
1969	5.0	5.1	4.0	9.9	7.0	8.3	31.0	37.9
1970	4.9	5.2	4.2	9.9	6.9	8.2	31.1	37.7
1971	5.1	5.2	4.1	9.3	6.8	8.4	30.7	37.2
1972	5.2	5.2	4.1	9.1	6.6	8.6	30.4	36.6
1973	5.2	5.2	4.2	8.9	6.1	8.6	30.4	36.5
1974	5.2	5.2	4.1	8.8	6.0	8.8	30.4	36.6
1975	5.2	5.2	4.2	8.7	6.1	8.8	30.4	36.6

Note: [a] Although all the other columns in this table exclude both enterprise employment and the armed forces, the last column *includes* both of these elements (and correspondingly adjusts total service employment to include the armed forces also).
Source: Appendix.

Another way to put Canadian bureaucratic growth into perspective is to consider briefly what has happened in other, presumably somewhat comparable, countries. International comparisons of any variety are always treacherous, and those of public employment ratios are no exception.

[6] Fuchs (1977) makes a similar point in the U.S. context and notes that causation indeed more likely runs the other way, that is, what growth in government employment there has been is in part at least a consequence of the growth of the service economy.

Nevertheless, Table 4 boldly compares the growth of public employment in Canada, the United States, and the United Kingdom over a recent 15-year period. Although the data are not strictly comparable either in coverage or definition, the differences do not seem great enough to vitiate the overall impression that the patterns and changes over time in these three countries are remarkably similar. Both at the beginning and the end of the period, the United States was lowest and the United Kingdom highest, even though public employment grew more rapidly in the United States than in the United Kingdom: Canada's public employment share actually grew most slowly of the three. Excluding government enterprises, in all three countries the subnational levels of government were by far the most important employers. Britain's central government was of course the largest of the three in relative terms, both because it has no states or provinces and because hospital workers are central government employees in Britain. The post-war nationalization of substantial sectors of British industry is also clearly reflected in the relative size of public enterprises—as is, to a lesser extent, Canada's long tradition of public enterprises. All three countries also show marked declines in the relative size of their military forces.

As Table 4 suggests, it appears to be a myth to assert that what has happened to public employment in Canada in recent years differs from what has happened in a number of other advanced industrial countries, namely, a very slow increase in what most consider to be an already large public employment share. Longer-term studies for various countries also show similarities to what appears to have happened here: in all cases, the greatest growth in public sector employment appears to have taken place during and immediately after the war, rather than in more recent years. Over the country as a whole, however, it appears to be a reality that the relative importance of government employment (a narrower concept) has increased—approximately doubled—in a number of countries.[7]

The broad cross-national similarity of these patterns suggests that the undoubted bureaucratic growth in Canada during the twentieth century probably reflects less the vices and virtues of the particular governments that happened to be in power than the broad sweep of change affecting the entire Western world in the post-war period. This is not to say that Canada exhibits no peculiarities or that government policy is irrelevant in the face of historical determinism. The point is rather that the political process in Canada appears to have functioned in a very similar way to that in other democracies in the face of rising populations, income, and expectations. In these circumstances, it is not at all difficult to understand why the public sector has

---

[7] This rather impressionistic assessment is based on the Canadian data presented earlier and on data for France in André and Delorme (1978); for Australia in Barnard, Butlin and Pincus (1977), and Harris (1975), for the United Kingdom in Abramovitz and Eliasberg (1953), and Jackson (1978); and for the United States in Abramovitz and Eliasberg (1953), Ginzberg (1965) and Hiestand (1976). It would be a major task to assemble the data in these and other sources in a really comparable form.

Table 4
PUBLIC EMPLOYMENT: AN INTERNATIONAL COMPARISON
(As per cent of total employment)

	Canada		U.S.A.		U.K.	
	1961	1975	1961	1975	1961	1975
Total Government and Government Enterprises[a]	22.2	23.8	18.9	20.6	22.4	27.4
Total National Government and Enterprises	7.5	5.7	8.8	6.9	14.6	15.9
Civilian, Government	3.3	3.3	3.1	2.7	3.4	6.4
Armed Forces	1.9	0.8	4.6	3.1	2.0	1.4
Government Enterprises	2.2	1.6	1.2	1.1	9.2	8.1
Total Subnational Government and Enterprises	14.7	18.0	10.1	13.7	7.8	11.5
Education	4.9	6.2	4.5	6.7	3.2	5.9
Other, Government	8.3	9.9	5.0	6.1	4.6	5.6
Government Enterprises[b]	1.6	1.9	0.6	0.9	—	—

Notes: [a] Columns may not add exactly, because of rounding.
[b] In the U.K. there are no separately identifiable government enterprises at the local level. Such enterprises are included at the central level of government.
Source: Jackson (1979).

expanded. In a sense, what is more surprising is that it has expanded so surprisingly little in terms of real resources, including labour, of which it makes direct use.[8]

EMPLOYMENT BY LEVEL OF GOVERNMENT

In 1975, 75 out of every 100 public employees in Canada worked, in essence, for provincial and local government. Moreover, over the post-war period as a whole, the relative importance of federal employment has been declining steadily, except for the case of civil service employment. Is there much variation in conclusions such as these when one looks at the data on a province-by-province basis? What is the geographic distribution of public employment? And how do the different provinces compare? These are the sorts of questions considered briefly in this section.

According to taxation data, in 1975 Quebec and Ontario together accounted for 61 per cent of public employment in Canada—apparently up from 1947 but about the same proportion as in 1961 (Foot and Thadaney 1978). The four Atlantic provinces together in 1975 accounted for 9 per cent, or less than British Columbia alone (11 per cent), while the three Prairie provinces accounted for 19 per cent. The most noticeable trend over the post-war period has been the rise in the proportion of total public employees

[8] In real terms, the share of national output consumed directly by government has hardly altered in 20 years; this broader picture will be treated in more detail in a separate paper.

located in Quebec, from 12 per cent in 1947 to 20 per cent in 1975, although Ontario still remained by far the most important area in terms of the sheer number of public employees located there.

Interestingly, increased *federal* employment accounted for only 11 per cent of the increase in public employment in Quebec over this period (compared to 16 per cent in Ontario and 20 per cent in Nova Scotia): the bulk of the increase in Quebec was thus clearly due to factors other than the relocation of federal offices to Hull. Close to two-thirds of federal employees were in the two central provinces in 1975, as in 1947—although there appears to have been some limited decentralization to the Atlantic and Prairie provinces, particularly during the 1950s. Public sector employment as a whole was also somewhat concentrated in the central provinces than was total employment, in contrast to the apparent situation in 1947.

This finding perhaps lends some support to the frequent complaints about the 'undue' geographic concentration of the federal bureaucracy (see Fullerton 1977). The considerable extent to which the nature of federal functions—other than delivering the mail and collecting taxes—limits the need for direct federal contact with citizens throughout the country must also be remembered, however. Most direct 'service' functions are subnational in Canada, and this is of course reflected in the employment data.

A clearer picture of the changing trends in the provincial distribution of public employment is provided in Table 5, which shows that there were, in total, almost twice as many public employees relative to total population in 1975 as in 1961, and almost three times as many as in 1947. The fastest increase per capita terms was in the Maritime provinces, all of which by 1975 were close to the national average, in marked contrast to earlier years. Manitoba had the highest per capita level of public employment in 1975 (apart from the exceptional case of the Northern Territories), displacing British Columbia, the leader in 1947 and 1961. The West as a whole appears high on this scale, while Ontario was actually below average in 1975, and Quebec, despite its striking rise in the 1960s was still, with Newfoundland, at the bottom of the list.

In terms of growth rates, a quite different provincial ranking can be constructed from the taxation data. This time Newfoundland and Quebec (with P.E.I.) are at the top of the league, with the highest growth rates in the post-war period, while Manitoba and, especially, Saskatchewan are at the bottom. Federal employment has grown particularly slowly in these two provinces. Indeed, as suggested above, federal employment growth on the whole appears to have been relatively 'centralist' in recent years (with the exception of P.E.I. and Alberta). Provincial employment has also grown more slowly in Manitoba and especially Saskatchewan than in most other provinces—a reality which may surprise those who associate socialist governments with bureaucratic growth—as has municipal government. Ontario too is a 'low growth' province in these areas, although it has had an above-average growth in educational employment over the post-war period as a whole. In general, Saskatchewan stands out as showing below-average growth in all components of public employment—and Quebec as displaying above-average growth in all—though neither province, it will be recalled, stands at the extreme in terms of public employment per capita (Table 5).

Table 5
PER CAPITA PUBLIC EMPLOYMENT[a] BY PROVINCE, TAXABLE AND TOTAL NON-TAXABLE RETURNS
(Selected Years)

Province	1947	1961	1975
Newfoundland	—	27.1	75.3
Prince Edward Island	9.9	32.4	87.4
Nova Scotia	21.2	45.3	91.6
New Brunswick	16.5	37.8	82.4
Quebec	13.3	31.0	75.5
Ontario	31.5	54.8	82.8
Manitoba	25.5	53.4	107.3
Saskatchewan	22.5	48.0	84.5
Alberta	24.9	52.8	93.0
British Columbia	34.3	55.7	93.6
Yukon and N.W.T.	22.5	100.5	160.4
Total Taxable	24.3	46.1	84.6
Total Non-Taxable	9.9	9.5	16.4
Total Public	34.3	55.6	101.0

Note: [a] Returns per thousand of population.
Source: Foot and Thadaney (1978).

The spectacular growth in the absolute size of public sector employment in the post-war period was indeed accompanied by some change in the relative importance of public employment in different provinces, with the most striking increases being in the northern territories, some of the Atlantic provinces, and Quebec, and the most striking decline in Ontario. Nevertheless, at the end of the period, as at the beginning, Ontario remained the place with the most public employees, and public employment remained more important in proportion to population in the Western provinces than in the East, despite the generally lower rates of growth in the former (apart from Alberta). A more important conclusion, then, is that one should be very cautious in generalizing about such popular subjects as the effects on public employment of electing different political parties and the effects of federal decentralization policies. Although all the evidence is far from in, it should already be obvious that reality is much more complicated than is normally thought and that the drawing up of unidimensional 'league tables'—who's first? . . . who's second? . . . and so on—is not a particularly productive activity.

The Statistics Canada data reported in Foot, Scicluna and Thadaney (1978b) provide some additional information on the provincial distribution of government employment (a narrower concept than that discussed above). These data too show that the Atlantic provinces all have above-average per capita employment, as do the Western provinces (excluding Manitoba). The two largest provinces—Quebec and Ontario—although accounting for almost three-fifths of total provincial government employment throughout the period, both have well below-average employment per capita. On the whole, the general pattern of distribution of provincial government employment by province shown in these data too has changed relatively little over the last decade.

Much of the same is true of municipal employment, where again Ontario and Quebec—especially the former—have consistently accounted for over 70 per cent of the total for municipalities with over 50,000 residents over the last decade. Once again, facile generalizations seem dangerous: although the Ontario share has been rising while the Quebec share has been falling, Ontario does not have the highest growth rate nor does Quebec have the lowest. The Atlantic provinces have all experienced below-average growth in municipal employment while the Western provinces (except Saskatchewan) have all had above-average growth. Apart from New Brunswick and Saskatchewan, the available data suggest that the distribution of local government employment by province is close to that of provincial government employment.

EMPLOYMENT BY FUNCTION

Data on the functional distribution of government employment are available for all three levels of government for recent years. Although there are some differences in definition and coverage, Table 6 presents a brief summary picture of the functional distribution of government civilian employment at the end of 1975. It is clear from this table that most *government* employees continue to be engaged in the traditional 'hard' functions of government: as noted earlier, the situation is, however, quite different if one considers the distribution of total *public* employees. The fact is, nevertheless, that activities as familiar to Adam Smith as general administration, protection, transportation, and public works accounted for about 52 per cent of all government employees in 1975 (a proportion which would have risen to 60 per cent if the armed forces were included). The concentration at the provincial level of government of the 'soft' functions—health, education, welfare, and recreation—which have accounted for so much of the expansion in spending in the post-war era is reflected in employment patterns, since these functions account for half of all provincial employees in 1975. On the whole, however, even though most teachers and hospital employees are excluded from these data, thus substantially understating the importance of education and health, the continued dominance of the traditional functions of government in terms of employment is noteworthy.

Table 6
FUNCTIONAL DISTRIBUTION OF GOVERNMENT CIVILIAN EMPLOYMENT, 1975
(Per Cent)

	Federal	Provincial[a]	Municipal[b]	Total[c]
General Government	19.6	10.3	13.3	14.6
Protection	22.9	14.0	30.4	20.5
Transportation, Works[d]	26.4	13.1	24.8	20.5
Health	2.2	21.0	3.0	10.2
Education	1.0	20.1	—	8.8
Social Welfare	11.6	7.1	6.5	8.8
Recreation, Culture	2.4	2.1	21.1	5.7
Other	13.9[e]	12.4[f]	0.9[g]	10.9
	100.0	100.0	100.0	100.0

Notes: [a] Excludes British Columbia.
[b] For municipalities with over 50,000 residents only.
[c] Weighted average: note that exclusion of British Columbia provincial employment and employment in municipalities with less than 50,000 population affects results. If *all* municipal employment is assumed to be distributed as shown above, the major change in the 'total' column would be to raise slightly the percentage shares of protection, works, and especially recreation, and lower the others.
[d] 'Transportation and communications' for federal (including Post Office) and provincial governments; public works (16.8); sanitation (4.0); and water works (4.0) for municipal governments.
[e] Includes natural resources and environment (4.6); immigration and citizenship (1.0); agriculture, industry and trade (4.4); foreign affairs (2.1); and miscellaneous (1.8).
[f] Includes local development (1.1); natural resources (4.9); agriculture, trade and industry (4.5); and miscellaneous (1.9).
[g] Includes miscellaneous (0.9).
Source: Foot, Scicluna, and Thadaney (1978b).

Functional data for federal civilian employment from 1957 to 1975 show that the concentration of federal employment in the basic activities mentioned above has hardly altered over the last two decades. Although, as might be expected in view of the sharp decline in the absolute size of the armed forces, only 23 per cent of federal civilian employees were classified as 'protection' in 1975 (this category includes civilian defence employees, police, correctional services, and courts) compared to 34 per cent in 1957, this decline was almost completely offset by the increase in general government employees and, especially, employees in transportation and communications (notably the Post Office). Outside the armed forces (79,817 in 1975), the largest federal employers were the Post Office (59,236) and the Department of National Defence (38,425). Although the absolute number of civilian defence employees has been shrinking fairly steadily for decades (there were 55,017 such employees in 1957 for example), this decline has been more than offset by the rise in federal police and correctional employees (from 8,836 to 27,645) over the same period. At least in terms of federal employment, it seems more accurate to refer to the growth of the 'police state' than the 'welfare state'!

Excluding the declining defence function, federal employment grew on average at 4.2 per cent annually from 1957 to 1975; including defence, the annual growth rate was only 2.9 per cent, with growth being below trend for most of the 1960s and (as noted earlier) relatively faster in the early 1970s. By function, the fastest growth rates were in some of the minor activities (notably foreign affairs), though health, transportation and communications, recreation and culture, education and social welfare all had above-average growth rates. In short, as one might expect, it is a reality that employment generally grew most quickly in those 'social' areas where federal expenditure was expanding most rapidly; but it is a myth to think that these areas account for much of federal employment (17 per cent in 1975, or 14 percent, including the armed forces).

The provincial data are unfortunately less complete in that Quebec is excluded before 1966 and British Columbia is excluded almost completely. Nevertheless, it seems safe to conclude that the effects of changing expenditure patterns on provincial employment have been more marked than at other levels of government. Although the proportion of provincial employees in health has, perhaps surprisingly, remained fairly constant (about 20 per cent) over the last decade, the proportion in welfare and, especially, education has risen sharply (from 10 per cent in 1961 to 27 per cent in 1975). Interestingly, fewer provincial employees are classified as 'general government' than at the other levels of government, and this proportion (about 10 per cent) has remained fairly stable. From 1961 to 1976 provincial employment grew at an average annual rate of 4.8 per cent, with the fastest growing functions being local development, education (even adjusted for New Brunswick's provincialization of education), and social welfare. After the first half of the 1960s, however, the growth rate of education declined sharply to well below average, although those for welfare, local development, and health increased.

Considering the sources of growth by province, a number of interesting features emerge: for example, all of the Atlantic provinces (and Manitoba) recorded above-average growth in general movement; Quebec has had the highest growth in health, Saskatchewan the lowest; for education, again the Atlantic provinces generally lead, with Ontario having the lowest rate. Across all functions, however, Saskatchewan, though a relatively high employer in per capita terms, had by far the lowest growth rates, with the Atlantic provinces (apart from New Brunswick) and the other Prairie provinces at the high end. Faced with such confusing results, plausible generalizations are, as noted earlier, hard to come by.

Finally, municipal data on the functional distribution of employment are still less comprehensive, since they cover only the 1967-1976 period and exclude all municipalities with fewer than 50,000 people (which, for example, leaves P.E.I. out completely). As shown in Table 6, the most important employers in Canadian municipalities are, as one might perhaps have expected, the police and fire departments, followed by parks and public works departments. By far the fastest-growing function in terms of employment over the last decade, however, has been social welfare, followed by general government and recreation. In general, it appears that the Atlantic provinces—which have experienced below-average growth in municipal employment—have above-average shares of that employment in the protection function and below-average shares in recreation (as does Quebec). On the whole, however, the general functional distribution of municipal employment across provinces is very similar—indeed, amazingly so considering the tremendous variation in the relative importance of (large) municipalities as employers, ranging from only 4 per cent provincial employment in New Brunswick and 5 per cent in Newfoundland to 20 per cent in Saskatchewan, 35 per cent in Quebec, 40 per cent in Manitoba, and a high of 68 per cent in Ontario.

WOMEN IN THE PUBLIC SECTOR

In recent years there has been a good deal of interest in the opportunities afforded women in public sector employment (Cook 1976). Unfortunately, the only available data bearing on this question are preliminary and for the federal government only (Foot, Scicluna and Thadaney 1978b).

At first glance, these data suggest that the federal government has indeed become more of an 'equal opportunity' employer. From 1961 to 1975, for example, the female proportion of federal government employment rose from 26 to 33 per cent, with most of this increase taking place after 1970 (when the female share was 29 per cent). In other words, over the 1961-1975 period as a whole, 45 per cent of those added to the federal civilian work force were women. This increase appears, however, to be no more than in line with the increase in labour force participation by women over these years (Cook 1976, p. 117). Furthermore, by far the greatest concentration of female employees in the federal government continues to be in the 'soft' functions—traditionally female-dominated—of health, social welfare, and education. In 1975, these were the only functions in which women constituted a majority of federal employees.

Even more interesting are the occupational data, which suggest that most women employed by the federal government continue to be office workers: indeed, the increase in the already marked female dominance of the 'administrative support' category in this short period—66 to 77 per cent—is striking. At the other end of the occupational scale, although the rate of increase in female executives—particularly at the very end of the period—has been striking, the absolute numbers involved are small and, in total, only 25 per cent of the increment in female employment over this period occurred in the three 'top' occupational classes,[9] compared to 66 per cent of the increment in male employment. That is, two out of every three men added to the federal work force in the early 1970s were 'professionals' (executive, scientific and professional, or administrative and foreign service) compared to only one in four women. The fact that this differential greatly exceeds the male-female differential enrolment in full-time post-secondary education in the late 1960s and early 1970s (Cook 1976, p. 57) perhaps suggests women still have some distance to go, at least in the federal public service.

OCCUPATIONAL STRUCTURE

The only broad-based data on the occupational structure of public sector employment in Canada comes from the census (which, as noted earlier, employs a rather narrow concept of such employment). This source shows that in 1971, 29 per cent of all those employed in government were classified as 'clerical' and another 35 per cent in 'service' occupations. At the other end of the occupational scale 12 per cent were classified as 'managerial' and 11 per cent in various 'professional' categories. These proportions varied somewhat by level of government: the federal government (excluding defence) employed the greatest proportion of managers and clerks, followed by the provincial and local governments, respectively, while provincial governments employed the greatest proportion of 'professionals' and local governments the least. More detail than this is available only for the federal government and only for a few recent years (Foot, Scicluna and Thadaney 1978b).

Two occupational groups—administrative support (clerical) and operational—have accounted for well over half of federal government employment in recent years. More interesting, perhaps, there has clearly been a much faster rate of growth in the administrative and foreign service, and in the small executive group, than in other occupational categories. The popular view of a rapidly expanding upper echelon (though still small in absolute terms) in federal government is thus quite correct—though these figures alone can tell us nothing about the desirability or otherwise of this change.

In any case, it seems clear that substantial 'professionalization' has taken place in the federal government in the last few years; 48 per cent of the federal employees added between 1969 and 1975 fell into one of the three top categories, and the 'professional' share of the total rose from 16 to 23 per

---

[9] Executive, scientific and professional, and administrative and foreign service.

cent. To put it another way, for every additional office employee (administrative support) hired during this period, the federal government added 1.4 employees in the 'professional' grades—indeed, one in the administrative and foreign service category alone. While federal employment as a whole rose by 31 per cent from 1969 to the end of 1975, the number of 'scientific and professional' employees rose 72 per cent, the number of 'administrative and foreign service' employees rose 105 per cent, and the small 'executive' class increased in size by 142 per cent.

The federal government has thus, it appears, become a much more 'professional' organization in a relatively brief period, a development perhaps to be expected in the light of the increased sophistication and complexity of modern times, and hence of governmental tasks. Indeed, if the public sector is broadly interpreted to include health and education personnel, that sector now employs over two-thirds of all 'professionals'—broadly, those with a post-secondary education—in Canada, although only 5 to 10 per cent of all professionals are directly employed by governments (Gunderson 1979). Many of the remaining one-third, though nominally employed in the private sector, are very likely dependent at least in part upon the public sector for their employment (e.g., engineers working on government contracts or lawyers negotiating with government regulators or tax officials). In other words, almost the entire population of Canada with more than secondary school education is dependent, directly or indirectly, upon the public sector for their livelihood (as well as their education). Few facts about public employment seem potentially more significant, though it would take us well beyond our assigned task to explore further here the implications of the dual role of the public sector as supplier and demander of highly educated manpower.

CONCLUSION

This paper has covered a vast statistical territory—and is itself in large part only a summary of much more detailed work reported elsewhere. Even persistent and assiduous readers may readily be forgiven if they are reeling by this time from all the numbers. It may therefore be useful to conclude by summarizing in point form the principal conclusions emerging from this discussion—categorizing each as 'myth' or 'reality' as it conflicts with or confirms what appear to be the dominant public perceptions about the nature of bureaucratic growth in Canada.

Reality —'The public sector is a large employer':
over 2.2 million people in Canada—24 per cent of the employed labour force—worked directly for public sector employers in 1975.

Myth —'The federal government accounts for most public employment':
less than one quarter of total public employees were federal and provincial civil servants—and less than one quarter in total worked

for the federal government in any capacity. In contrast, provincial governments employed 44 per cent of this total (over two-fifths of them in hospitals alone).

Myth —'Public employment growth has been especially fast in recent years':
the fastest relative growth in public employment was probably in the early post-war period. Public employment as a whole grew only very slowly in relative terms after 1960 and has remained almost constant since the early 1970s.

Reality —'The civil service has expanded rapidly recently':
the federal civil service grew particularly quickly in the early 1970s and provincial civil services have grown at a high, steady rate for the last two decades.

Myth —'Public sector growth is turning Canada into a 'service economy':
although public employment rose slightly as a proportion of total employment since 1960, it has declined significantly as a proportion of service sector employment. Four out of five new jobs created in Canada in recent years have been in the service sector but only one in four has been in the public sector—and only two in a hundred in the federal government.

Myth —'Canada's rate of bureaucratic growth is abnormally high':
trends in public employment development in Canada appear to be roughly comparable to those in the United States, the United Kingdom, and other industrial countries.

Reality —'Most public employees are located in central Canada':
most public employees are located in Ontario and, to a lesser extent, Quebec. The same is true of federal employees. This distribution has changed little in the post-war period. On the other hand, there are more public employees *per capita* in the West than in the East, and the fastest increases (in these terms) in the post-war years have been in the Eastern provinces—and the fewest provincial and municipal employees per capita are in Ontario and Quebec.

Myth —'Most government employees are in social services':
most *government* employees continue to be engaged in such traditional government functions as general administration, protection, transportation, and public works. It is, however, a *reality* that most *public* employees are in social services.

Myth —'Affirmative action is alive and well in Ottawa':
while the federal government is indeed employing relatively more women, most of them continue to be employed as administrative support staff and in the traditionally female-dominated social service areas.

Reality —'Most government employees are bureaucratic office workers': most *government* employees (as opposed to public employees more broadly defined) are clerks and administrators, but an increasing number are 'professionals'. Indeed, the public sector as a whole (including health and education) employs two-thirds of all Canadians with post-secondary education.

APPENDIX

ESTIMATED PUBLIC SECTOR EMPLOYMENT, 1961-1975

An earlier paper (Bird 1978) presented a comprehensive estimate of public sector employment, broadly defined, for 1975. The present paper presents similar estimates of each year of the 15-year period 1961 to 1975.[10] The basic estimates are presented in Table A1. These numbers are explained column by column in this appendix, as are the supplemental data assembled in Table A2.

TABLE A1

Federal civil service:
From Table A2.5, Bird (1978); originally from *Annual Reports* of the Public Service Commission. Increase in 1967 reflects extension of 'public service' concept in that year. In 1973, 2,000 employees in agencies not previously recorded were included; in 1974, 2,353 employees not previously under the *Public Service Employment Act* were included for the first time.

Other federal civilian:
'Federal employment' from Foot, Scicluna, and Thadaney (1978b) *less* 'federal civil service' in previous column. Average for end-quarter figures. This column obviously reflects the definitional changes noted in the comments on 'federal civil service'.

Armed Forces:
From Department of National Defence. Strength of the regular forces as of March 31.

Provincial civil service:
From Table A2.5, Bird (1978); originally from *Annual Reports* of the provincial Public Service Commissions, from Hodgetts and Dwivedi (1974), or from direct correspondence with provincial Public Service Commissions. There are various differences over time and among provinces with regard to the scope of these data: for details see notes to Table A2.5 in Bird (1978). In particular, in Newfoundland, hospital employees are included in 'civil service' figures.

Other provincial:
Average end-quarter 'provincial employment' from Foot, Scicluna, and Thadaney (1978b) adjusted to include data missing for Quebec for 1961-1965 and for British Columbia since 1962 *less* 'provincial civil service' in previous column. Quebec data for 1961-1965 were estimated by assuming the rate of growth over this period was the same as that for total provincial employment recorded in Foot, Scicluna, and Thadanay (1978b): this assumption is valid for example, for the 1966-1968 period. For British Columbia, total provincial government employment for 1975 was estimated as follows: in 1961 Statistics Canada reported 17,282 non-enterprise employees in British Columbia, compared to 10,233 employees in the B.C. public service in that year (data from the B.C. Government Employee Relations Bureau). In 1975, the latter source reported 39,301 public service employees: this figure is not directly comparable with the 1961 figure, however, for some enterprise and casual employees are now included. Adjusting for the approximate effect of

[10] The 1975 data in the present appendix differ slightly from those in the earlier study owing largely to the fact that most of the figures in the earlier estimate were year-end figures.

including these classes of employees (by comparing pre-inclusion 1965 and post-inclusion 1966 figures), adjusted 'public service' employment in 1975 was estimated at 22,402, or 219 per cent greater than in 1961. The total number of provincial employees in 1975 was then calculated at 37,848 by applying this percentage to the 1961 Statistics Canada figure. It was then assumed that in each year between 1961 and 1975—the two end-points for which figures were available—employment grew by the same absolute amount. Although this method is obviously crude, the numbers involved are so small that further refinement of an already shaky estimate was judged not worthwhile. The 'other provincial' figures include certain educational and hospital employees: see later comments.

Municipal:
Average end-quarter 'municipal employment' from Foot, Scicluna, and Thadaney (1978b) and average end-month figures from Statistics Canada, *Municipal Government Employment, 1961-66* (72-505).[11]

Education:
This series embodies more estimation than any other in this table. The only firm figures are those for 1972-1975, which are annual average monthly data from the Labour Division of Statistics Canada: see Table A2.2 in Bird (1978). The 1972 data are partly estimated because information on vocational employees was available only for the last six months of the year. The figures for 1961-1971 inclusive have been estimated on the basis of Statistics Canada Education Division data on full-time teachers only, that is, excluding both 'casual' teachers and non-teaching staff, as well as all vocational employees. For the 1972-1975 period for which direct comparisons between these two series can be made, it is clear that there was a marked increase in the ratio of non-teaching staff, casual teaching staff, and vocational staff to full-time non-vocational teachers. Nevertheless, for all earlier years the full-time teacher-figure has simply been 'blown up' by the ratio for 1972. Although it seems likely that this procedure results in an overestimate, it was decided not to extrapolate back the observed change in the ratio in 1972-1975 because the results quickly become absurd in that all non-teaching staff virtually vanishes after a few years. The extreme shakiness of the resulting estimates needs to be underlined, however, as education throughout the period (in these figures) remained the largest single employer in the public sector.

Two other problems with the education figures also deserve notice. In the first place, these figures include a relatively small number of employees of private schools. The result is that both educational employment and total public sector employment are slightly overstated. It should also be noted that throughout the period all universities and colleges are counted as part of the public (or 'para-public') sector regardless of their actual legal status. This treatment seems justified in view of the overwhelmingly public nature of financial support for post-secondary education.

The second problem concerns the possible overlap between these figures and those reported elsewhere as 'provincial employment'. This overlap was resolved by reducing the 'education' total by all employees included under the heading "Education" in Statistics Canada, *Provincial Government Employment* (72-007). Data for Quebec 'Education' employees were estimated for 1961-1965 on the assumption that this number grew at the same rate as total provincial employees (see comments above under 'other provincial'). No adjustment was made for British Columbia (or for the relatively few federally employed teachers). The 'education' data in Table 1 thus *exclude* a significant number—65,000 in 1975, for example—of 'education' employees who are included in the 'other provincial' column. This procedure accounts for a large proportion of the changes noted in that

---

[11] All references to Statistics Canada publications include the catalogue number (in parentheses).

column: in 1971, for example, all New Brunswick secondary school teachers became provincial employees. Even more important, in 1973 there was an increase of over 20,000 in the employees of post-secondary educational institutions included in the provincial government data for Ontario and Quebec. As in the case of the 'civil service' data, it is important to distinguish changes such as these, which reflect changes in statistical definitions.

Hospitals:
From Table A2.3 in Bird (1978) and originally from data supplied by the Health Division of Statistics Canada. The 1975 figure is partly estimated on the assumption that total hospital employment grew at the same rate (5.0 per cent) as non-medical employment in general and allied special hospitals. These data, like the education data, are slightly overestimated because they include employees of proprietary hospitals—less than 1 per cent of total in 1974—and treat all hospitals as 'public' in the sense of being under the financial and operational control of government. Again as with the education data, double-counting was reduced by excluding from these figures all employees listed under 'hospital care' in Statistics Canada, *Provincial Government Employment* (72-007). No adjustments were made for missing Quebec and B.C. data (or for federal hospital employees). The noticeable drop in the estimate in Table 1 for 1973 is due to the inclusion in the provincial employment series for that year, for the first time, of about 15,000 hospital workers in Alberta and Quebec.

Federal enterprises:
From Foot, Scicluna and Thadaney (1978b). Includes employees of enterprises and of administrative, regulatory, and special funds.

Provincial enterprises:
From Statistics Canada, *Provincial Government Employment* (72-007). Includes enterprises and special funds. British Columbia enterprise employment for 1962-1971 inclusive was estimated by assuming that employment grew annually by the same absolute amount. Quebec enterprise employment for 1961-1965 was estimated as having grown at the same rate as the total for the provincial enterprises shown in the source.

Municipal enterprises:
This series is based on an employment series for urban transit obtained from Canadian Urban Transit Association, *1975-6 Transit Fact Book*, plus a crudely estimated adjustment for other municipal enterprise employees based on the ratio between transit employees and others estimated for 1975 in Bird (1978). The basis for this series is perhaps even shakier than that for 'education', but the numbers are small.

Total:
Sum of the preceding columns.

As per cent of Service Employment:
'Service employment' includes transportation, communications, and other utilities; trade; finance, insurance, real estate; community business and personal service; public administration. Data are from Statistics Canada, *Historical Labour Force Statistics* (71-201) and are averages of month-end totals. The 'armed forces' data in Table 1 have been added to 'service employment' for purposes of this calculation.

As per cent of Total Employment:
From Statistics Canada, *Historical Labour Force Statistics* (71-201): excludes residents of Territories, Indians, and inmates of institutions. The 'armed forces' data in Table 1 have been added to 'total employment' for purposes of this calculation.

TABLE A2

Federal civilian employment is the sum of 'federal civil service' and 'other federal civilian' in Table 1.

Total federal government is the sum of 'federal civilian employment' and 'armed forces' from Table 1.

Provincial is the sum of 'provincial civil service' and 'other provincial' in Table 1.

Total civil service is the sum of 'federal civil service' and 'provincial civil service' in Table 1.

Total civilian government is the sum of 'federal civilian', 'provincial', and 'municipal' from Table 1.

Total government is 'total civilian government' **plus** 'armed forces'.

Broad provincial is 'provincial' **plus** 'hospitals' from Table 1 **plus** estimate of the number of educational employees at the post-secondary level not already included in 'provincial'. The estimate for 1961-1971 is based on the relationship for 1972-1975 between the data reported in Table A2.3 of Bird (1978) and the 'education' figures in Statistics Canada, *Provincial Government Employment* (72-007).

Broad municipal is 'municipal' from Table 1 **plus** those education employees **not** included in 'broad provincial'.

Total enterprises is the sum of federal, provincial, and municipal enterprises from Table 1.

Total education is 'education' from Table 1 **plus** those education employees included in 'provincial' in Table 1.

Total hospitals is 'hospitals' from Table 1 **plus** those hospital employees included in 'provincial' in Table 1.

Total para-public is the sum of 'total education' and 'total hospitals'.

Total civilian public is 'total' from Table 1 **less** 'armed forces'.

Total non-enterprise civilian public is 'total civilian public' **less** 'total enterprises'.

Total federal is 'total federal government' **plus** 'federal enterprises' from Table 1.

Total provincial is 'broad provincial' **plus** 'provincial enterprises' from Table 1.

Total municipal is 'broad municipal' **plus** 'municipal enterprises' from Table 1.

Table A1
ESTIMATED TOTAL PUBLIC SECTOR EMPLOYMENT, 1961-1975
(in thousands)

Year	Federal Civil Service	Other Federal Civilian	Armed Forces	Provincial Civil Service	Other Provincial	Municipal	Education	Hospitals	Federal Enterprises	Provincial Enterprises	Municipal Enterprises	Total	As Percent Service Employment	As Percent Total Employment
1961	135.9	68.5	120.0	100.7	67.7	140.8	291.1	210.4	138.8	67.9	29.9	1371.7	40.4	22.2
1962	137.3	66.4	126.5	106.3	65.0	145.0	310.7	222.2	137.3	67.6	29.8	1414.1	40.2	22.2
1963	137.8	64.8	123.7	113.1	67.5	149.6	324.5	242.3	135.0	69.3	29.6	1457.2	39.2	22.3
1964	138.7	63.7	120.8	125.1	61.1	154.5	343.4	259.1	137.6	70.3	29.9	1504.2	39.1	22.3
1965	140.2	67.6	114.2	134.5	62.6	159.8	356.7	275.2	138.6	75.0	31.3	1555.7	38.4	22.2
1966	145.8	74.5	107.5	158.9	48.7	166.3	383.7	289.8	142.5	77.3	33.0	1628.0	38.1	22.4
1967	200.3	31.1	105.7	174.1	47.1	175.4	410.7	313.4	150.4	85.2	34.6	1728.0	38.4	23.0
1968	200.3	34.5	101.7	185.6	50.2	184.0	435.5	325.5	138.4	89.4	34.8	1779.9	38.2	23.2
1969	199.7	37.9	98.3	192.8	50.4	191.9	468.7	332.8	140.9	91.3	34.9	1839.6	37.9	23.3
1970	198.7	42.8	93.4	202.8	51.9	203.7	484.3	335.7	138.9	93.2	34.6	1880.0	37.7	23.6
1971	216.5	43.5	89.6	211.1	66.4	208.2	473.3	342.9	137.6	96.9	35.3	1921.3	37.2	23.4
1972	230.8	45.8	84.9	226.5	63.6	215.6	480.6	347.1	137.9	100.0	36.7	1970.4	36.6	23.3
1973	245.3	45.5	82.4	234.1	100.1	232.0	496.7	337.7	142.8	107.8	40.8	2065.2	36.5	23.2
1974	258.6	45.1	81.8	247.6	113.6	239.3	511.7	349.6	149.0	116.7	44.3	2157.3	36.6	23.4
1975	273.2	41.1	79.8	259.9	115.1	251.3	525.1	369.5	148.3	132.6	47.2	2243.1	36.6	23.7
% Change 1961-1975	101	−40	−34	157	70	78	80	76	7	95	58	64		

Sources and Notes: See Appendix.

Table A2
SUPPLEMENTARY DATA ON PUBLIC EMPLOYMENT, 1961-1975
(in thousands)

Year	Federal Civilian	Total Federal Government	Provincial	Total Civil Service	Total Civilian Government	Total Government	Broad Provincial	Broad Municipal	Total Enterprises	Total Education	Total Hospitals	Total Para-Public	Total Civilian Public	Total Non-Enterprise Civilian Public	Totals		
															Federal	Provincial	Municipal
1961	204.4	324.4	168.4	236.6	513.6	633.6	418.0	392.7	236.6	300.9	234.6	535.5	1251.7	1015.1	463.2	485.9	422.6
1962	203.7	330.2	171.3	243.6	520.0	646.5	439.9	409.3	234.7	322.3	247.3	569.6	1287.6	1052.9	467.5	507.5	439.1
1963	202.6	326.3	180.6	250.9	532.8	656.5	472.1	424.9	233.9	336.8	268.6	605.4	1333.5	1099.6	461.3	541.4	454.5
1964	202.4	323.2	186.2	263.8	543.1	663.9	498.1	445.1	237.8	356.6	286.6	643.2	1383.4	1145.6	460.8	568.4	475.0
1965	207.8	322.0	197.1	274.7	564.7	678.9	530.3	458.5	244.9	371.2	304.1	675.3	1441.5	1196.6	460.6	605.3	489.8
1966	220.3	327.8	207.6	304.7	594.2	701.7	559.8	487.6	252.8	399.3	319.5	718.8	1520.5	1267.7	470.3	637.1	520.6
1967	231.4	337.1	221.2	374.4	628.0	733.7	617.8	502.9	270.2	431.5	341.9	773.4	1622.3	1352.1	487.5	703.0	537.5
1968	234.8	336.5	235.8	385.9	654.6	756.3	650.0	530.3	262.6	457.8	356.0	813.8	1678.2	1415.6	474.9	739.4	565.1
1969	237.6	335.9	243.2	392.5	672.7	771.0	652.0	584.6	267.1	487.7	367.6	855.3	1741.3	1474.3	476.8	743.3	619.5
1970	241.5	334.9	254.7	401.5	699.9	793.3	662.8	615.6	266.7	502.4	373.5	875.9	1786.6	1519.9	473.8	756.0	650.2
1971	260.0	349.6	277.5	427.6	745.7	835.3	738.0	563.9	269.8	502.7	380.0	882.7	1831.7	1561.9	487.2	834.9	599.2
1972	276.6	361.5	290.1	457.3	782.3	867.2	767.8	565.6	274.6	510.5	384.4	895.4	1885.5	1610.0	499.4	867.8	602.3
1973	290.8	373.2	334.2	479.4	857.0	939.4	782.3	618.3	291.4	553.6	391.3	944.9	1982.8	1691.4	516.0	890.1	659.1
1974	303.7	385.5	361.2	506.2	904.2	986.0	798.9	635.9	310.0	573.8	905.8	979.6	2075.5	1765.5	534.5	915.6	680.2
1975	314.3	394.1	375.0	533.1	940.6	1020.4	863.4	657.5	328.1	589.7	426.3	1016.0	2163.3	1835.2	542.4	996.0	704.7
% Change 1961-1975	53.8	21.5	122.7	125.3	83.1	61.0	106.6	67.4	38.7	86.0	81.7	89.7	72.8	80.8	17.1	105.0	66.8

Sources and Notes: See Appendix.

Chapter Nine

The Other Side of Public Spending: Tax Expenditures in Canada

by
*Allan M. Maslove**

The government of Canada currently spends close to $50 billion annually in budgetary expenditures. The estimates of these expenditures are laboriously produced by the federal bureaucracy, scrutinized and approved by Parliament, and the expenditures themselves are examined by the Auditor General. This occurs every year. The government also "spends" an amount, probably in excess of $8 billion per year,[1] in a manner that almost entirely escapes any form of governmental or public scrutiny. The items that enter into this latter total, for the most part, receive initial authorization by Parliament and then rarely are examined again unless they are amended in some fashion. The actual dollar amounts involved are never authorized in any direct vote. These items have come to be referred to as tax expenditures.

Tax expenditures can be defined briefly as special provisions in the tax laws providing for preferential treatment and consequently resulting in revenue losses.[2] They are designed for one or both of two purposes. The first purpose is to grant tax relief to individuals in special circumstances, for example, people incurring unusually high medical costs. The second purpose is to provide individuals or corporations with special incentives to pursue certain courses of action. The tax expenditures considered in this paper are provided for either in the form of deductions from taxable income or in the form of credits that reduce tax liability,[3] although exclusions from taxable income could also be regarded as tax expenditures.

These special tax provisions are expenditures in the sense of being

---

* I wish to thank Irwin Gillespie and Rick Van Loon for their helpful comments on an earlier draft of this paper. Chris Hannah served as a diligent research assistant and editor.

[1] This total is based on estimates of corporate tax expenditures of $2.4 billion in 1973 (Perry 1976) and personal tax expenditures of $5.5 billion in 1976 (derived later in this paper).

[2] The standard personal deduction could be considered as a tax expenditure, although none of the studies cited in this paper view it in this way.

[3] It is of considerable importance whether one chooses to deliver tax expenditures by means of deductions rather than credits. Deductions, because they are adjustments to the tax base, result in tax relief directly related to the taxpayer's marginal tax rate. Tax credits grant the same dollar amount of relief to all taxpayers, irrespective of their marginal rates. Thus, while the two methods can be designed to produce the same tax reduction in the aggregate, their distributional consequences will differ.

revenues the government chooses not to collect; hence the term "tax expenditures." The purposes to which tax expenditures are directed could be pursued by means of budgetary expenditures. Thus, the questions to be raised in the analysis of tax expenditures concern their allocational, stabilizing, and distributive effects as compared to budgetary expenditures or other instruments in pursuing policy goals.[4]

The purpose of this paper is to provide an introduction to some of the issues involved in the tax expenditure concept and to undertake, at a fairly aggregative level, some preliminary analysis of the impact of personal tax expenditures in Canada. Attention is restricted mainly to the personal income tax system largely for reasons of data availability; personal tax expenditures constitute over 50 per cent of the total. On the corporate side, the estimates of Perry (1976) will be reviewed.

In the following sections, some of the more prominant methodological issues will be discussed, and estimates of personal tax expenditures in Canada will be presented and analysed. A final section will outline the beginnings of a theory of tax expenditures and suggest two proposals for policy reform.

METHODOLOGICAL ISSUES

Obviously the issue one must begin with is: what is a tax expenditure? What constitutes a special deduction or credit under the tax law as distinct from the normal structure of the system? The definition of the normal tax structure in turn requires a definition of taxable income.

Beginning with theoretical principles, the concept of income perhaps most often adopted in the tax expenditure literature, and in public finance generally, is known as the Haig-Simons definition. Income is defined broadly as the total accretion to wealth or the change in net worth plus consumption over the accounting period. This concept is, in fact, very close to that adopted by the Carter Commission on Taxation,[5] though it is considerably broader than that adopted by modern income tax systems. In addition, economists would generally designate the family as the income earning unit, while in the Canadian tax laws it is the individual.

Because of the divergence between the economist's "ideal" income tax base and that actually employed, analysts of tax expenditures have also adopted a second guide-line based on historical practice: tax expenditures constitute departures from the normal or accepted structure of the income tax.

---

[4] The reasonableness of the goals themselves involve broader questions and will not be raised here. Rather, the position taken here is, given the policy goals, are tax expenditures an effective means of attainment?

[5] *Report of the Royal Commission on Taxation* (Ottawa 1966). The comprehensive tax base is defined as "the sum of the market value of goods and services consumed or given away in the taxation year by the tax unit, plus the annual change in the market value of the assets held by the unit." For administrative purposes it is redefined as the sum of a series of "net gains, appropriately defined, or each tax unit on an annual basis." (Vol. 3, pp. 39-42). The relevant "net gain" items are listed.

This second criterion is not without problems. For example, the basic personal exemption and progressive marginal tax rates are generally considered to be part of the normal structure, although one could certainly argue the opposite. Other items are even more debatable. For example, should the failure to tax the imputed incomes that homeowners receive from their dwellings be viewed as a tax expenditure? According to the Haig-Simons income definition the answer is clearly yes; however, imputed incomes from owner-occupied dwellings do not enter the base of most tax systems and on this basis the general practice has been not to include them in tax expenditure analyses.

While the definitional issue casts some uncertainty on the analysis of tax expenditures, two other methodological issues are narrower and more technical. The first involves the computational procedure. Because of the form of the data, it is necessary to adopt a procedure in which the revenue cost of each tax expenditure is calculated on the assumption that the remainder of the tax system is fixed. However, if the removal of one or more of these provisions boosts the taxpayer into a higher marginal tax bracket, the resulting estimated revenue losses will be biased downwards.

The second point is the assumption that economic behaviour is unchanged. In a general equilibrium context, however, it is likely that in response to the removal of one or several tax expenditures, taxpayers would adjust their income earning and spending behaviour. Unfortunately, beyond the statement that these adjustments will be attempts to offset the increased tax liabilities, it is difficult to be any more precise about their nature. In particular, we do not know whether the estimated amounts will be over or underestimates due to the direct effects of these individual adjustments, nor how the resultant changes in general economic activity will indirectly further affect the tax expenditure estimates. Overall then, one cannot easily predict whether behavioural adjustments will mean the tax expenditure estimate will be upward or downward biased.

DEVELOPMENTS IN THE U.S.

The *Congressional Budget Act* of 1974 required that the budget presented annually to Congress by the president include a listing of tax expenditures. Credit for this development has, in large part, been attributed to the strong advocacy of Stanley S. Surrey, formerly an assistant secretary in the U.S. Treasury Department and now a professor of law at Harvard University.[6] His book *Pathways to Tax Reform: The Concept of Tax Expenditures* (Surrey 1973) is one of the most comprehensive studies in the field.

Given this new exposure, one might expect some impact on future tax

---

[6] See, for example, Carl S. Shoup (1974).

policy in the United States. At this point in time, it is probably too early to discern any clear patterns in the data. However, a comparison between Canada and the United States later in this paper does provide some interesting preliminary results. Table 1 shows the amounts of tax expenditures over the last several years and their relationship to budgetary revenues and expenditures. Tax expenditures in the United States have grown at an annual rate of about 9 per cent over the years 1974-1978 (between 1975 and 1976 the growth rate was 17 per cent). Corporate tax expenditures in the United States amount to approximately 24 per cent of the total, while in Canada the split between corporate and personal tax expenditures appears to be more even. Tax expenditures have declined slightly relative to budgetary expenditures since 1974, but there is no trend relative to budgetary revenues.

Table 1
FEDERAL TAX EXPENDITURES IN THE UNITED STATES

Year	Corporate Tax Expenditures ($ million)	Personal Tax Expenditures ($ million)	Total Tax Expenditures ($ million)	Total as % of Budgetary Revenues	Total as % of Budgetary Expenditures
1972	13,350	46,460	59,810	28.7	25.8
1974	17,465	57,340	74,805	28.2	27.9
1975	19,295	62,110	81,405	29.0	25.1
1976	22,935	72,425	95,360	31.8	26.0
1977	25,005	78,810	103,815	29.1	25.8
1978	26,505	86,210	112,715	28.2	24.4

Source: Calculated from the Budget of the United States Government, Special Analyses and S. S. Surrey, Panel Discussion on the Budgetary Process.

TAX EXPENDITURES IN CANADA

We turn now to estimates of tax expenditures in Canada. The analysis in this paper is restricted to the personal income tax system, but before getting to that a summary of Perry's estimates of corporate tax expenditures is presented. Perry estimated a corporate tax expenditure total of $2,368 million in 1973. The largest single item was due to the difference between capital cost allowance claimed and true depreciation. Other items of major importance were due to exploration, development, and depletion allowances and non-taxable capital gains. By industry, the biggest beneficiaries (measured in terms of tax concessions as a proportion of taxes paid) were agriculture, petroleum, and coal. When measured in dollar amounts the major beneficiary industries were mining, petroleum and coal, and financial services.

There are two studies of personal tax expenditures in Canada that have been published to date. The first, entitled *The Hidden Welfare System*, is by the National Council of Welfare (N.C.W.) (1976). This publication, as the title suggests, strongly casts the tax expenditure system in terms of a hidden welfare system for the rich that in dollar size is much bigger than the explicit welfare system for the poor. Seventeen out of 60 tax expenditure items are quantified amounting to a total of $6.4 billion for 1974.[7] This amount is substantially larger than that reported later in this paper; the reason is the inclusion of several additional items in the National Council of Welfare study. Unfortunately insufficient data are provided to calculate a total comparable to the total in this paper.

The N.C.W. study also finds a heavy bias in favour of higher income taxpayers. Those receiving total incomes under $5,000 in 1974 received an estimated average benefit of $244, while those with incomes in excess of $50,000 received $3,990. The lowest 70 per cent of taxpayers received 40 per cent of the benefits while the highest 11 per cent received 33 per cent. Among the largest of the individual tax expenditure items were interest income deductions ($546 m.), registered retirement savings plan contributions ($513 m.), and registered pension plan contributions ($467 m.).

The second study by Kesselman[8] adopts a narrower view of what constitutes a tax expenditure. He estimates a tax expenditure total of $2.2 billion for 1973. Kesselman also finds the benefits are distributed strongly in favour of higher income groups with registered pension plan contributions, registered retirement savings plan contributions, and charitable donations being among the most unequal individual items. In terms of cost, the registered pension ($273 m.) and retirement savings plans ($270 m.) were the largest.[9] In addition to the effect of making the personal income tax system less progressive, Kesselman estimates that if tax expenditures were eliminated, average tax rates could have been 19.5 per cent lower.

The estimates presented in this paper are more aggregative than those of Kesselman or the National Council of Welfare. Rather than focus on individual tax expenditure items, this paper concentrates on developments over time and equity effects using total estimates. The estimating procedure is thus less detailed than Kesselman's but for the year in common (1973) the

---

[7] The seventeen quantified items in the N.C.W. study are listed below. Those included in the present study are indicated with the notation "(p.s.)":

age exemption, disability exemption, married or equivalent exemption, dependent children exemption, other dependents exemption, education deduction, deduction for CPP/QPP contributions (p.s.), unemployment insurance premiums (p.s.), registered pension plan contributions (p.s.), retirement savings plan premiums (p.s.), registered home ownership plan contributions (p.s.), interest income (p.s.), tuition fees (p.s.), child care expenses (p.s.), charitable donations (p.s.), the standard deduction (p.s.), and the federal tax reduction provision in effect in 1974. A full list of items included in this study is provided in the Appendix.

[8] Kesselman (1977). The tax expenditure items included by Kesselman are the same as those in the present study (listed in the Appendix). In addition, Kesselman includes one other item, education expenses, which in dollar terms is relatively minor.

[9] The other large items, the Registered Home Ownership Savings Plan and the interest income deduction, did not exist in 1973, the year of Kesselman's estimates.

estimates of the totals are very close (Kesselman estimates $2.2 billion while the estimates of this paper are $2.4 billion).

The estimates presented are deliberately downward biased for two reasons. First is the point raised earlier, namely, that as deductions and exemptions are removed, taxpayers will move into higher marginal tax brackets. The second reason is that a conservative view of what constitutes a tax expenditure is adopted; many other items could arguably be included. For example, the results in this paper are based on sixteen items (see the Appendix) while the National Council of Welfare claims to have identified approximately 60 separate items, of which seventeen were estimated.

Tables 2 through 4 summarize the pattern of personal tax expenditures in Canada over the last several years. Between 1971 and 1976 total tax expenditures grew at an average annual rate of about 32 per cent, substantially faster than either budgetary revenues or expenditures. As a consequence, tax expenditures as a portion of federal revenues grew from 6 per cent to 15 per cent, and as a portion of direct expenditures from 6 per cent to over 14 per cent.[10] While it does not appear that any strong trends in growth rates by income group exist, there is a clear distributive pattern in favour of higher income groups. Beginning from about $3,000, as annual total income rises, tax expenditures as a percentage of income steadily

Table 2
PERSONAL TAX EXPENDITURES IN CANADA

Year	Personal Tax Expenditures ($ million)	% of Budgetary Revenues[1]	% of Budgetary Expenditures[1]
1967	653.3	6.1	6.0
1971	1,372.4	8.1	8.1
1972	1,956.8	10.1	10.0
1973	2,435.5	10.8	11.2
1974	3,694.9	12.5	13.2
1975	4,387.7	13.9	12.8
1976	5,458.0	15.4	14.5

Source: Calculated from Appendix Tables A-1 to A-7 and *Government Finance* (Statistics Canada, 68.001).
Note: 1. Revenues and Expenditures are on a National Accounts Basis to make them compatible on a calendar year basis with the tax expenditure estimates.

[10] The tax expenditures are estimated by a formula which includes an allowance for the provincial tax share (see the Appendix). Nevertheless, it is assumed that the entire revenue loss is from the federal treasury. This assumption implies that the federal-provincial agreements resulting in federal tax point abatement would be altered in the absence of these tax expenditures to maintain aggregate provincial income tax revenues at roughly the same dollar amount.

increase. In dollar terms, in 1976 the average tax expenditure received by a taxpayer reporting income under $5,000 was $75, between $15,000 and $20,000 of income it was $694, and for those reporting more than $50,000 of income it was $6,613.

In addition to vertical equity, the presence of tax expenditures raises the question of horizontal equity. That is, taxpayers at a given income level may pay different amounts of income tax because of differing abilities to take advantage of tax deductions. To get a clear picture of the extent of this effect one would ideally like to construct a matrix relating total income to either

Table 3
PERSONAL TAX EXPENDITURES PERCENTAGE GROWTH RATES
1971-1976

Growth Rates Between	Total Tax Expenditures	Average Tax Expenditures	Average Tax Expenditure by Selected Income Group			
			under $5,000	$10,000-$15,000	$20,000-$25,000	over $50,000
1971-72	42.6	30.6	51.2	5.0	-0.5	6.1
1972-73	24.5	17.6	-4.8	2.1	0.1	19.4
1973-74	51.7	43.9	30.5	20.2	11.6	17.7
1974-75	18.8	15.1	2.6	-0.7	-11.0	-10.4
1975-76	24.4	20.8	-0.1	0.1	-0.0	18.0
1971-76	297.7	206.9	82.9	35.5	-0.0	57.6

Source: Calculated from Appendix Tables A-1 to A-7.

Table 4
TAX EXPENDITURES AS A PERCENTAGE OF TOTAL INCOME

INCOME GROUP	YEAR						
	1967	1971	1972	1973	1974	1975	1976
under $1,000	2.8	4.2	4.9	4.8	4.6	3.6	2.5
$1,000-$2,000	1.2	2.2	2.8	2.6	2.6	1.9	1.2
$2,000-$3,000	0.8	1.6	2.2	2.0	2.5	2.0	1.2
$3,000-$4,000	0.9	1.3	2.2	2.2	3.3	3.7	3.6
$4,000-$5,000	1.2	1.5	2.3	2.2	3.0	3.5	3.7
$5,000-$10,000	1.6	1.8	2.4	2.4	3.0	3.3	3.3
$10,000-$15,000	2.3	2.7	2.8	2.8	3.4	3.3	3.5
$15,000-$20,000	3.3	4.1	4.0	4.0	4.5	3.9	4.0
$20,000-$25,000	3.9	5.1	5.0	5.1	5.6	5.0	5.0
$25,000-$50,000	4.1	5.5	5.5	5.9	6.0	6.1	6.5
$50,000 and over	4.9	5.7	5.7	6.7	7.7	6.8	8.2

Source: Tables A-1 to A-7.

taxable income or taxes paid. The data available for this paper were not sufficient for building this matrix. Instead, one must rely on less informative data for some indication of horizontal inequities. Tables 5 and 6 indicate the percentage of tax filers in 1975 who paid no tax by level of income. Table 5 shows that about one-half of 1 per cent of tax filers (about 22,000

Table 5
TAXABLE AND NON-TAXABLE RETURNS BY INCOME GROUP
1975

INCOME GROUP	% OF TOTAL RETURNS THAT ARE:		AVERAGE TAX OF TAXABLE RETURNS (Dollars)
	TAXABLE	NON-TAXABLE	
under $1,000	1.2	98.8	25
$1,000-$2,000	0.7	99.3	17
$2,000-$3,000	13.7	86.3	19
$3,000-$4,000	27.7	72.3	70
$4,000-$5,000	63.1	36.9	179
$5,000-$10,000	91.6	8.4	684
$10,000-$15,000	99.4	0.6	1,699
$15,000-$20,000	99.6	0.4	2,885
$20,000-$25,000	99.6	0.4	4,163
$25,000-$50,000	99.6	0.4	7,297
$50,000 and over	99.5	0.5	27,260
TOTAL	70.8	29.2	1,834

Source: *Taxation Statistics*, 1977 edition.

Table 6
PERCENTAGE OF TAX FILERS PAYING NO TAX BY OCCUPATION AND INCOME GROUP
1975

OCCUPATION	INCOME GROUP					
	under $2,000	$2,000-$3,000	$3,000-$4,000	$4,000-$5,000	$ 5,000-$10,000	over $10,000
Employees	98.8	82.3	66.1	23.4	3.0	0.3
Farmers	99.3	93.3	86.0	71.7	37.3	1.9
Fishermen	99.5	48.8	41.5	28.3	7.0	0.3
Professionals	97.3	84.1	72.7	43.4	13.6	0.3
Salesmen	98.9	80.0	72.7	31.0	10.6	0.4
Business Proprietors	99.4	85.3	73.4	48.9	19.2	0.4
Investors	99.5	96.9	93.3	78.8	35.1	5.1
Pensioners	99.8	99.5	95.8	91.2	49.4	0.9
TOTAL	99.2	86.3	72.3	36.9	8.4	0.5

Source: *Taxation Statistics*, 1977 edition.

individuals) with incomes over $10,000 paid no tax in 1975. There also appears to be substantial variation by occupation (Table 6), reflecting the differing abilities of individuals in differing circumstances (but with similar incomes) to take advantage of deductions available to them.

ANALYSIS OF THE ESTIMATES

To summarize, the estimates presented in the previous section reveal a number of patterns.

1) Tax expenditures in Canada have been growing relative to budgetary expenditures and revenues, while in the United States the ratios have been relatively constant.[11]
2) Tax expenditures are distributed heavily in favour of higher income groups, thus decreasing the progressivity of the personal income tax system.
3) Tax expenditures appear to result in fairly significant horizontal inequities within income groups.

Given these patterns one might be tempted to dismiss tax expenditures as simply a fancy term for tax loopholes. However, as noted at the outset of this paper, they merit examination because they are directed towards public policy goals which may be pursued by the use of alternative instruments. In this light, one might ask whether the tax expenditure instrument does have some advantages. For example, aid to charitable institutions could be granted directly by the government. However, by allowing deductions from taxable income instead, the allocation of funds to charity is determined in a more decentralized fashion. The administrative costs are thus likely to be lower and more importantly, the allocation of funds, because it is determined by individuals, will more accurately reflect citizen preferences as to the relative merits of the recipient institutions. Nevertheless, it is still true that the present system means that all other taxpayers subsidize the donor's gift and the higher the donor's income the greater is the subsidy.

Other tax expenditure items allow for the adjustment of incomes for factors that result in expenditures but are often not considered part of the normal utility-producing basket of consumption goods and services. The deduction for medical expenses is an example. However, in this case again the value of the deduction is directly related to the taxpayer's income.

Other items would appear to have even less to recommend them. The Registered Retirement Savings Plan deduction is intended to encourage saving for retirement. However, the extent to which the RRSP actually does encourage new net saving as opposed to simply providing a tax-deferring diversion for saving that would have occurred in any event is debatable.

---

[11] This is true of the United States whether one looks at total tax expenditures (Table 1) or only personal tax expenditures.

The Registered Home Ownership Savings Plan is another example. It is designed to assist potential homeowners in the purchase of a dwelling. Again, one could ask whether the RHOSP is only a vehicle for diverting savings to avoid taxation. On the other hand, if it actually does encourage new net saving, it is debatable whether the RHOSP is a very cost-efficient method of assisting families in the purchase of homes.[12]

On balance, I think one must adopt a highly skeptical view concerning the efficacy of tax expenditures as instruments of public policy. Why, then, does the government appear to rely on them so heavily? In the next section, some directions in which a possible answer may lie are suggested.

TOWARDS A THEORY OF TAX EXPENDITURES

One approach to explaining the existence and growth of tax expenditures involves the choice of policy instruments by governments. In addition to choosing among potential policy goals, one can view the political (and bureaucratic) decision maker choosing among policy instruments to attain a given set of goals. There exists a "market" in policy instruments with decision makers choosing those that have the lowest associated costs (broadly defined).[13] Tax expenditures are thus very attractive candidates because their effects and costs are largely hidden, or at any rate not subject to nearly the same level of scrutiny as many other instruments. Direct expenditures are probably the most likely alternative in this context. One could then hypothesize that as the relative political costs of direct expenditures versus tax expenditures altered, the degree of reliance on each of these instruments would vary inversely.

The trends reported in Table 2 are interesting in this context. Over the 1970s as unemployment and inflation rates continued to rise together, public resistance to more government spending was increasingly articulated and included demands for actual spending cutbacks.[14] The political costs of the direct expenditure instrument thus increased over the course of this period. Accordingly, political decision makers relied relatively more heavily on alternative instruments, one of which was tax expenditures.[15] Table 2 shows that tax expenditures relative to budgetary revenues and expenditures grew steadily over the period.

A comparison with the trend in the United States between 1974 and 1978 (Table 1) is of further interest. The *Congressional Budget Act* of 1974, by

---

[12] Another, more fundamental, question of goals is left aside, namely, should the government be encouraging and subsidizing dwelling purchases at all?

[13] This concept is examined in Phidd and Doern (1978), chapters 2 and 14.

[14] Increasing resistance to higher taxes would, of course, accompany the opposition to more expenditures.

[15] Regulation is another alternative and one could also observe an increasing reliance on a variety of regulatory forms, with the imposition of wage and price controls perhaps being the outstanding and most extreme example.

requiring an annual tax expenditure statement, raised the political cost of this instrument relative to others and relative to what it had been previously. It did so by requiring the president to make explicit what had previously been largely hidden from public view. Accordingly, over a time period and in an economic climate similar to the Canadian, we find that tax expenditures did not grow in relative terms. The data for both the Canadian and American experience are inadequate for the performance of any reliable statistical tests, but they are nonetheless strongly suggestive.

Another (complementary) avenue of explanation flows from the distributive effects of tax expenditures. Table 4 (as well as the studies of Kesselman and the National Council of Welfare) indicates a consistent pattern over time in favour of higher income groups. The pattern in the United States is similar.[16] It is difficult to avoid the conclusion that this diluting of the progressivity of the income tax is systematic.

The ultimate explanation of this pattern is likely to be related to a more fundamental theory of the political economy of income redistribution, a theory which we do not possess. There may, however, be some validity in the notion of an ''equilibrium'' distribution of income and that the institutions of government, including the income tax system, are modified by equilibrating forces in a manner that produces this net result. Clearly this is an area requiring a great deal of further research.

TWO PROPOSALS

I conclude this paper with two proposals which together or individually would constitute modest reforms of the public economy.

Proposal I The minister of Finance should be required to produce and make public an annual tax expenditure statement.

This proposal simply involves the adoption of a statement similar to that in the budget of the United States since 1974. The statement should be prepared as part of the documentation provided by the Finance Department in support of the minister's budget speech.

The proposal is based on the simple premise that more public information results in better public policy. In terms of the discussion in the previous section, an annual tax expenditure statement would increase the political costs of using this set of policy instruments and may thus have some effect on public decision making. It is also recognized that if there is something like an equilibrium income distribution, an annual statement may result in other hidden changes elsewhere that would have offsetting effects. One thinks of the analogy of a balloon that is squeezed in one place only to bulge in another.

---

[16] See, for example, Surrey (1973), chapter 3.

The immediate response of the opponents of an annual statement will be that the concept of tax expenditures is not yet sufficiently clear and that doubt exists regarding what constitutes a tax expenditure. Indeed, this is true and the early parts of this paper recognized this point. However, there exists a sufficient degree of concensus to make such a statement clearly meaningful. A comparison with GNP, the National Accounts, unemployment rates and price indices is not inappropriate. These data have been produced for decades and areas of uncertainty still remain in each of them. Moreover, the official adoption of a tax expenditure statement would itself be a spur to further developmental work.

Proposal II: Each tax expenditure item should be written into the *Income Tax Act* for a period of not more than five years. If not re-enacted after that time the measure would automatically lapse.

This "sunset law" provision is designed to force a periodic explicit reconsideration of each tax expenditure item. Many would probably receive virtually automatic re-endorsement if their purposes were deemed to be still valid. An allowance for medical expenses may be an example. Others, though, would likely be closely examined and some measures, determined to be ineffective or whose earlier goals had lapsed, would be allowed to expire. The RHOSP may be an example of the latter.

Since these measures would not all come up for reconsideration at the same time, they would not likely overload the system or create a new "analysis industry." However, if this were perceived to be a problem, the sunset measure could be applied only to those tax expenditures whose cost to the federal treasury was greater than some threshold level.

CONCLUSION

In addition to seriously diluting the progressivity of the personal income tax system, tax expenditures constitute a major drain on its yield. In 1976, for example, the estimated revenue lost through tax expenditures was 31 per cent of personal income taxes actually collected.

The potential for redesign of federal government taxes and expenditures if these preferential measures were to be removed is substantial. Thus, expenditure evaluation exercises that deal only with budgetary expenditures are too narrow in scope; both sides of the public purse merit examination.

APPENDIX

THE ESTIMATION OF TAX EXPENDITURES

The data upon which the estimates are based are taken from the appropriate editions of *Taxation Statistics* published by Revenue Canada. Table 2 of these publications is the major source. The total deductions from income form the basis for the estimation of tax expenditures. These deductions included CPP or QPP contributions, Unemployment Insurance premiums, Registered Pension Plan contributions, Retirement Savings Plan premiums, Registered Home Ownership Plan contributions (beginning in 1974), interest and dividend income deduction (beginning in 1974), pension income deduction, union and professional dues, tuition fees, child care expenses, general expense allowance, other employment expenses, standard deductions, medical claims, charitable donations and other deductions.

For each income group the average taxable income per taxpayer was calculated and from this an applicable marginal tax rate was determined. This was adjusted upward to include the provincial portion of the income tax by using the tax share of Ontario. Ontario was selected because approximately 40 per cent of all taxpayers reside in this province and because the Ontario tax rate is at an intermediate level. The resulting marginal tax rate for each income group was then multiplied by the total deductions of that group to yield the estimate of tax expenditures for the group.

The results along with other related data and estimates are reported in Tables A-1 to A-7.

Table A-1
ESTIMATES OF TAX EXPENDITURES
PERSONAL INCOME TAX SYSTEM
1967

INCOME GROUP	Percentage of Total Tax Payers (%)	Total Tax Expenditures ($,000)	Percentage of Total Tax Expenditures (%)	Average Tax Expenditure/ Taxpayer ($)	Average Income/ Taxpayer ($)	Tax Expenditures as % of Average Income (%)
under $1,000	11.2	10,896	1.7	12	435	2.8
$1,000-$2,000	13.2	18,900	2.9	18	1,502	1.2
$2,000-$3,000	13.3	22,556	3.5	21	2,502	0.8
$3,000-$4,000	13.3	35,996	5.5	33	3,488	1.0
$4,000-$5,000	12.2	52,091	8.0	53	4,491	1.2
$5,000-$10,000	30.7	262,775	40.2	105	6,754	1.6
$10,000-$15,000	4.1	90,927	13.9	271	11,821	2.3
$15,000-$20,000	1.0	46,323	7.1	567	17,068	3.3
$20,000-$25,000	0.4	27,873	4.3	870	22,183	3.9
$25,000-$50,000	0.5	53,318	8.2	1,342	32,960	4.1
$50,000 and over	0.1	31,646	4.8	3,768	76,660	4.9
TOTAL	100.0[1]	653,300	100.0	80	4,652	1.7

[1] Percentages will not total to 100% because individuals declaring zero or negative incomes are not included in the ''under $1000'' category.

Table A-2
ESTIMATES OF TAX EXPENDITURES
PERSONAL INCOME TAX SYSTEM
1971

INCOME GROUP	Percentage of Total Tax Payers (%)	Total Tax Expenditures ($,000)	Percentage of Total Tax Expenditures (%)	Average Tax Expenditure/ Taxpayer ($)	Average Income/ Taxpayer ($)	Tax Expenditures as % of Average Income (%)
under $1,000	9.0	20,013	1.5	23	553	4.2
$1,000-$2,000	12.0	38,014	2.8	33	1,502	2.2
$2,000-$3,000	10.4	39,768	2.9	40	2,485	1.6
$3,000-$4,000	10.0	43,688	3.2	46	3,496	1.3
$4,000-$5,000	8.9	56,298	4.1	66	4,491	1.5
$5,000-$10,000	34.0	418,132	30.5	129	7,220	1.8
$10,000-$15,000	10.3	311,039	22.7	318	11,847	2.7
$15,000-$20,000	2.3	151,297	11.0	688	16,962	4.1
$20,000-$25,000	0.7	78,462	5.7	1,120	22,109	5.1
$25,000-$50,000	0.8	137,524	10.0	1,808	33,080	5.5
$50,000 and over	0.2	78,139	5.7	4,196	74,231	5.7
TOTAL	100.00[1]	1,372,375	100.0	144	5,876	2.5

[1] See note, Table A-1.

Table A-3
ESTIMATES OF TAX EXPENDITURES
PERSONAL INCOME TAX SYSTEM
1972

INCOME GROUP	Percentage of Total Tax Payers (%)	Total Tax Expenditures ($,000)	Percentage of Total Tax Expenditures (%)	Average Tax Expenditure/ Taxpayer ($)	Average Income/ Taxpayer ($)	Tax Expenditures as % of Average Income (%)
under $1,000	7.6	23,044	1.2	29	595	4.9
$1,000-$2,000	10.9	47,513	2.4	42	1,494	2.8
$2,000-$3,000	9.9	56,337	2.9	55	2,492	2.2
$3,000-$4,000	9.8	77,077	3.9	76	3,501	2.2
$4,000-$5,000	9.3	100,197	5.1	103	4,491	2.3
$5,000-$10,000	33.9	609,034	31.1	173	7,269	2.4
$10,000-$15,000	12.6	435,464	22.3	334	11,922	2.8
$15,000-$20,000	3.0	206,847	10.6	673	16,968	4.0
$20,000-$25,000	1.0	111,719	5.7	1,114	22,116	5.0
$25,000-$50,000	1.0	179,428	9.2	1,793	32,910	5.5
$50,000 and over	0.2	110,165	5.6	4,452	77,599	5.7
TOTAL	100.0[1]	1,956,825	100.0	188	6,381	3.0

[1] See note, Table A-1.

Table A-4
ESTIMATES OF TAX EXPENDITURES
PERSONAL INCOME TAX STSTEM
1973

INCOME GROUP	Percentage of Total Tax Payers (%)	Total Tax Expenditures ($,000)	Percentage of Total Tax Expenditures (%)	Average Tax Expenditure/ Taxpayer ($)	Average Income/ Taxpayer ($)	Tax Expenditures as % of Average Income (%)
under $1,000	5.8	17,325	0.7	27	565	4.8
$1,000-$2,000	10.5	44,042	1.8	38	1,483	2.6
$2,000-$3,000	9.3	50,582	2.1	50	2,489	2.0
$3,000-$4,000	8.8	75,222	3.1	77	3,500	2.2
$4,000-$5,000	8.8	94,173	3.9	97	4,496	2.2
$5,000-$10,000	33.4	640,646	26.3	174	7,294	2.4
$10,000-$15,000	15.2	572,407	23.5	341	12,023	2.8
$15,000-$20,000	4.1	310,375	12.7	682	16,985	4.0
$20,000-$25,000	1.4	165,104	6.8	1,115	22,079	5.1
$25,000-$50,000	1.3	279,111	11.5	1,945	32,855	5.9
$50,000 and over	0.3	186,478	7.7	5,316	79,960	6.7
TOTAL	100.0[1]	2,435,465	100.0	221	7,066	3.1

[1] See note, Table A-1.

Table A-5
ESTIMATES OF EXPENDITURES
PERSONAL INCOME TAX SYSTEM
1974

INCOME GROUP	Percentage of Total Tax Payers (%)	Total Tax Expenditures ($,000)	Percentage of Total Tax Expenditures (%)	Average Tax Expenditure/ Taxpayer ($)	Average Income/ Taxpayer ($)	Tax Expenditures as % of Average Income (%)
under $1,000	5.1	15,639	0.4	26	575	4.8
$1,000-$2,000	9.1	40,866	1.1	39	1,508	2.6
$2,000-$3,000	8.2	59,699	1.6	63	2,494	2.5
$3,000-$4,000	7.7	102,270	2.8	115	3,497	3.3
$4,000-$5,000	7.5	117,885	3.2	135	4,502	3.0
$5,000-$10,000	31.5	800,157	21.7	219	7,321	3.0
$10,000-$15,000	18.7	887,457	24.0	410	12,167	3.4
$15,000-$20,000	6.8	593,505	16.1	757	17,019	4.5
$20,000-$25,000	2.2	322,196	8.7	1,244	22,052	5.6
$25,000-$50,000	1.9	440,972	11.9	1,957	32,526	6.0
$50,000 and over	0.4	314,216	8.5	6,255	80,788	7.7
TOTAL	100.0[1]	3,694,861	100.0	318	8,170	3.9

[1] See note, Table A-1.

Table A-6
ESTIMATES OF TAX EXPENDITURES
PERSONAL INCOME TAX SYSTEM
1975

INCOME GROUP	Percentage of Total Tax Payers (%)	Total Tax Expenditures ($,000)	Percentage of Total Tax Expenditures (%)	Average Tax Expenditure/ Taxpayer ($)	Average Income/ Taxpayer ($)	Tax Expenditures as % of Average Income (%)
under $1,000	4.1	10,276	0.2	21	569	3.6
$1,000-$2,000	7.8	26,791	0.6	29	1,530	1.9
$2,000-$3,000	7.2	42,482	1.0	49	2,498	2.0
$3,000-$4,000	6.9	105,925	2.4	125	3,490	3.7
$4,000-$5,000	6.5	124,049	2.8	159	4,504	3.5
$5,000-$10,000	30.3	873,016	20.0	240	7,364	3.3
$10,000-$15,000	20.1	984,989	22.5	407	12,271	3.3
$15,000-$20,000	9.3	743,130	16.9	669	17,116	3.9
$20,000-$25,000	3.4	447,289	10.2	1,107	22,088	5.0
$25,000-$50,000	2.8	651,322	14.8	1,969	32,174	6.1
$50,000 and over	0.6	378,418	8.6	5,602	81,975	6.8
TOTAL	100.0[1]	4,387,686	100.0	366	9,224	4.0

[1] See note, Table A-1.

Table A-7
ESTIMATES OF TAX EXPENDITURES
PERSONAL INCOME TAX SYSTEM
1976

INCOME GROUP	Percentage of Total Tax Payers (%)	Total Tax Expenditures ($,000)	Percentage of Total Tax Expenditures (%)	Average Tax Expenditure/ Taxpayer ($)	Average Income/ Taxpayer ($)	Tax Expenditures as % of Average Income (%)
under $1,000	3.4	6,052	0.1	14	561	2.5
$1,000-$2,000	6.5	15,334	0.3	19	1,553	1.2
$2,000-$3,000	6.8	26,311	0.5	31	2,498	1.2
$3,000-$4,000	6.5	99,800	1.8	125	3,494	3.6
$4,000-$5,000	5.9	123,044	2.3	168	4,496	3.7
$5,000-$10,000	27.7	842,026	15.4	246	7,410	3.3
$10,000-$15,000	20.4	1,086,620	19.9	431	12,364	3.5
$15,000-$20,000	11.8	1,012,067	18.5	694	17,169	4.0
$20,000-$25,000	5.1	684,218	12.5	1,095	22,106	5.0
$25,000-$50,000	4.2	1,050,588	19.2	2,049	31,627	6.5
$50,000 and over	0.6	511,984	9.4	6,613	80,389	8.2
TOTAL	100.0[1]	5,458,044	100.0	442	10,313	4.3

[1] See note, Table A-1.

Chapter Ten

Commentary

THE POLITICAL CONTEXT OF INEQUALITY IN A TIME OF CRISIS

by
Hugh Armstrong

My comments are structured around five observations. The first is the current series of cut-backs in spending in governments. For example, at the provincial level, Ontario grants to universities have been shrinking throughout the 1970s on a constant-dollar per student basis. This financial situation affects the faculty and support staffs, especially when enrolments level off and start declining. At the national level, one reflection of the cut-backs is the bitterness permeating relations between the federal government and its employees (e.g., the inside postal workers). In short, and except possibly in Alberta, things are getting tight in the state sector.

However, and this leads to my second observation, things are not equally tight for all. For one thing, somebody has to do the tightening. We seem to be the ones, whether as officials working directly for the state or as academics commenting and advising on this work. To use Professor Gillespie's term, we are among the technicians of government spending. We are not the "public." But if we are technicians, we are not and cannot do work that is exclusively technical. It is also necessarily *political*. There are no simple, neutral techniques of evaluating government spending, as Professor Gillespie demonstrates so carefully in his paper. Let me build on what he says about value preferences: they are not randomly distributed. For example, in arguing that government spending is out of control, some commentators focus not on the tax expenditures analysed by Professor Maslove, and not on the purchase of fighter planes, but on health and welfare, unemployment insurance benefits, the indexation of public service pensions, and Crown corporations. There is a good deal of government-bashing going on these days. It has been widely noted and occasionally practised here. My point is that it is *selective*. For example, we often hear of an Ottawa bureaucracy mushrooming out of control. Yet, from the *Taxation Statistics* data it can be seen that while in 1946 there were more federal state workers than provincial and municipal state workers combined, the provincial total passed the federal total by 1964 and the municipal total did so by 1967, with education and hospital workers excluded in all cases. The bashing is not indiscriminate. Put in more general terms, and this is my third observation, we technicians are not only necessarily engaged in political activity, but there is a pattern to the politics within which our work is situated.

My fourth observation is that the central feature of the patterns is inequality. Let me amplify this observation by picking up on some of the stimulating things in Mr. Renouf's paper. First, he uses a business analogy to suggest that we as the public view ourselves as "shareholders." We do not, however, have equal shares, the formal claims of citizenship notwithstanding. Second, he advocates viewing the public as "the Canadian community." This community is not to be seen as "simply a mass of individuals" but as a coherent entity. Yet, in historical, cultural, linguistic, and perceptual terms, and increasingly in territorial terms as well, we surely have at least two national communities in Canada. And it goes without saying that their positions are unequal. Moreover, I think it useful to consider that there are additional distinct communities that are based on income and wealth. Indeed, a case could be made for still others based on region and sex, and possibly even on age and race, although I would be prepared to argue that those based on income and especially on wealth are decisive, and embrace the others just listed. A third point mentioned by Mr. Renouf concerns the crisis of legitimacy facing the state. My comment here, and it is of course linked to the notion of 'publics' or 'communities' in the plural, is that different individuals and groups do not always agree on what is legitimate, and their views on legitimacy do not all have the same impact.

This leads to my fifth and final observation. The question of legitimacy relates both to who benefits and to who is perceived to benefit from state activities. Clearly the powerful do benefit. This is not simply a technical matter of analysing things like net fiscal incidence, important though such analyses are. Nor is it simply a matter of revealing ironies such as the $68 million grant from the federal and Ontario governments to Ford for a new engine plant being announced at about the same time as the cuts in the federal Consumer and Corporate Affairs grants program to consumer groups such as the Automobile Protection Association, and through it the Rusty Ford Owners' Association. At the most fundamental level, the powerful benefit from the fact that what the state seeks to preserve is not law and order in general, but a particular law and order, one which tends to favour the accumulation of capital. In the financial pages of the newspaper this function of the state is often referred to as creating an atmosphere of business confidence, or as establishing the conditions for economic growth. But it amounts to the same thing. The state helps the rich get richer.

The state is not, however, merely a tool in the hands of the capitalists, and here I return to the question of legitimacy. Given that our economic system depends on and promotes inequality, the state not only helps with the promotion, but it also helps cushion the effect of the promotion. Particularly now with the increased complexity of our society, the powerless are seen to benefit as well from the reforms embodied in the "welfare state."

So the state has the two, sometimes contradictory, functions of accumulation and legitimation, to use James O'Connor's terminology (although one must note the existence of a third, coercive state function as well). The accumulation and legitimation functions were both implicitly embraced by the federal government in its famous white paper of 1945, where it declared its public commitment to high and stable levels of both employment and income. The post-war boom saw this commitment apparently being met in Canada, as were similar commitments elsewhere in the advanced capitalist world, although our unemployment record was less satisfactory than most. But clearly the boom is now at an end. Instead, we seem to be in the throes of a long-term crisis. Again, Canada is a little worse off than most, but the crisis pervades the entire advanced capitalist world. The problems of government spending in bad times are faced throughout the West, and not just in Ottawa or across Canada. My fifth observation, then, is that the accumulation and legitimation functions of the state are increasingly difficult to fulfill and are increasingly in conflict as the world economic situation worsens.

To conclude, I have made five substantive observations. First, the immediate context of our deliberations is one of tightening in the state sector. Second, while we are discussing the 'public' evaluation of government spending, we are not the 'public' (and still less the 'publics'), but rather the technicians. Third, although we are technicians our work is not exclusively technical but also political, and furthermore, there is a pattern to the political context within which we operate. Fourth, the central feature of this pattern is inequality. Fifth, as the world economic crisis deepens, the accumulation and legitimation functions of the state become increasingly difficult to fulfill and come increasingly into conflict.

Let me hasten to add as a postscript that the outcome of this conflict is most certainly *not* pre-determined. What happens depends not only on technical matters associated with the logic of accumulation but also on the political activity of the politicians themselves, of technicians like us, and of the 'publics' out there. Government politicians in Ottawa and several of the provinces are of course already hard at it, for instance by blaming the victims—like women and young people—instead of the causes of unemployment.

PROGRAM EVALUATION: WHAT IS IT? WHO SHOULD DO IT?

by
James Cutt

Mr. Rogers' paper is important and provocative, containing many interesting and controversial points for discussion. I will just pull out a few general points from the paper for comment.

First, I would like to applaud Mr. Roger's courage in assuming his immensely challenging but very difficult mantle, and his view, firmly stated in his paper, that the cause for the evaluation of government spending and the development and implementation of evaluation procedures is strengthened, not weakened, by the very difficulties of the task. We have heard these pitfalls—or Pitfields as I believe they are sometimes called in Ottawa—eloquently expounded before, including some of the papers in this volume. At a time when the voices of hollow purism and trendy cynicism, especially in the academic world, often counsel despair, and when indeed the evidence seems to point to a very unimpressive track record for evaluation, Mr. Rogers is arguing that the task must and will be done, that accountability for the use of scarce public resources—defined across the spectrum from regularity and propriety through efficiency to effectiveness—requires a framework for the evaluation of public spending.

The role of the new Office of the Comptroller General in the development and implementation of evaluation in Canada is a particularly interesting one. The Treasury Board Secretariat's role, as I understand it, is to remain primarily in the *ex ante* evaluation of spending proposals and the making of recommendations to Treasury Board ministers on those proposals which will best achieve the government's objectives. On the other hand, the Auditor General's role is in the *ex post* evaluation of spending programs up to the level of efficiency auditing, with a mandate also to comment on the use of procedures for effectiveness auditing. The role of the new Comptroller General's office in evaluation is not to do evaluation at all, but to develop and ensure the implementation of procedures and processes, both on a routine and periodic basis, for the evaluation of efficiency and effectiveness in government departments. It thus has a procedures role which sits astride the process of evaluation as an ongoing, *ex ante*/monitoring/*ex post*, process. There is, of course, the other role in general financial management and administrative practices, and I should like to come back to it later.

On the particular role set out for the Comptroller General's office in evaluation, a few brief points might be made. First, it is surely ungainsayable that the role of establishing systems and procedures on a consistent basis which makes possible comparability is absolutely necessary. On the other hand, it would be possible to have a set of entirely consistent procedures which are nevertheless methodologically inadequate or outdated. So the question arises, as I understand it, whether the new office is to play a role of essentially auditing the use of evaluation procedures, or whether it is to assume the larger role of developing and criticizing methodology, and promoting or even insisting on the use of different procedures. In short, is the office to have a partially academic role, to be, in effect, a pool of technical expertise with a heavy research dimension? Mr. Rogers hints at this broader role but it might be interesting to pursue this point which would seem likely, in the Ottawa context, to have territorial implications.

Second, although the formal roles of the Treasury Board Secretariat, the Auditor General, and the Comptroller General seem clearly distinct, it is likely that a common view of the new office in departments will be that this is more paralysis by analysis, yet another mob of unproductive, quasi-academic trendies asking for data, diverting staff from the real work of pursuing government policy, and so on. There might therefore be much to be said for the co-ordination of the roles of the three evaluative components to prevent unnecessary duplication of data requests, and time wasting, and, perhaps more important, to ensure that the left hand knows what the right hand is doing methodologically. Mr. Rogers draws attention in his paper to the cost of evaluation itself, and one way to diminish the force of this criticism of the process is by apparent co-ordination of efforts.

Third, the mandate of the Comptroller General is explicitly on means, on evaluation instruments, and it will be a difficult, but important, task in the office to ensure that the focus of staff does not get diverted from the ultimate ends of those evaluative instruments. In an office concerned with procedures and methods, particularly if a quasi-academic role in the development and improvement of those methods emerges, it will be important but difficult to ensure that those means do not become ends in themselves. It is gratifying in this regard to have Mr. Rogers affirm in his paper his commitment to realism and practicality.

Fourth, on the actual methods to be pursued by the Office of the Comptroller General I would like to make three brief sub-points. First, far from the role of the Office of the Comptroller General in supervising financial management and related administrative procedures being in some sense inferior to or less rewarding than the broader role of the office in evaluation—and I do not argue for a moment that Mr. Rogers views the procedures in this way, but simply that the temptation will be there to opt for more glamourous activity—I believe that this first role is in itself of great importance, and, furthermore, forms the basis of a more sweeping role in evaluation. The establishment of the structural characteristics of a public information system by program and responsibility centre and the associated establishment of financial accounting procedures, in particular, the establishment of improved, detailed cost accounting procedures tied into control accounts in the general accounts of each department or agency, and the required use of accrual accounting across the board are, I believe, examples of procedures which are likely to offer a very high pay-off indeed.

Furthermore, such an information system forms the financial base and core of a broader information system, which brings me to my second sub-point.

Second, much of the failure, or what seems like failure, in evaluation, comes from aiming too high, at being too concerned with what Mr. Rogers describes as "big P" program evaluation and not concerned enough with "small p" program evaluation. My argument would be for starting small, establishing the core financial data base, and gradually aggregating upwards, rather than starting at a high and glamorous level of aggregation and attempting later, obviously with junior staff, to do the more pedestrian aggregation downwards. I would argue that microscopic relevance is better than macroscopic irrelevance, and that ultimate macroscopic relevance depends on a base of microscopic relevance. Or, if I may use a urological analogy, successful big P's depend on a lot of successful small p's. So much for depth. As to width, it again might be better to start small, perhaps on a sampling basis, and move gradually to a broader scope or width of coverage. There must again be a urological analogy on the usefulness of providing one adequate sample rather than a series of small samples, but perhaps we should not stretch the point too much.

Third, I would hope that in developing methodology, particularly in the area of output measurement, the Public Choice insistence that measures of success reflect consumer demand,[1] other than simply through the results of periodic elections, be borne in mind. Rod Dobell's story of the auditor whose objective was to successfully keep elephants off the Sparks Street Mall in Ottawa bears remembering. I would also dare to express hope that the new procedures are not dominated by economists, and that the analytical procedures recommended by the office include behavioural and institutional variables as well as the traditional techno-economic variables. I think we sometimes forget that government programs are implemented by people in organizations.

METHODS AND FORUMS OF EVALUATION: COMMENT

by
Pierre-Paul Proulx

My comments that follow are normative; they are not value-free. They comprise a shopping list of what one might look for in changing the process of public policy formulation and assessment. The general proposition from which I would start is that the public evaluation of government spending is a function of problems and priorities which places evaluation in a context of time and place, people and history. Within this specific context, the

---

[1] See, for example, Maslove (1975).

important problems themselves indicate the direction of desired change in the processes and forums for their evaluation.

Within this general framework, a number of basic policy problems suggest the need for modifications in the process of policy formulation and evaluation. First, economic space is not co-terminus with political space. There is growing interdependence among countries, for example, international trade, cartels, GATT and the structure of tariffs, and our relationship with the United States. These developments suggest that we need new institutions between government and the private sector and with other governments to adapt to these realities.

A second problem that calls for new forums for public evaluation of government decision making is that of dislocation and the need for adaptation to new environments. To adapt to the changing environment we shall have to develop methods which allow us to make decisions that discriminate between firms and sectors. A difficult situation becomes impossible if there are no alternative employment opportunities. This calls for the government to become a visible partner in international markets and, possibly, for the establishment of an agency, for industrial adaptation with guide-lines for partnership relationships between the public and private sectors. What we need is a framework to influence private decisions, not supplant them.

Third, in my view the pendulum has swung towards the provinces, and what is more important now is to examine forums for the interprovincial evaluation of policies rather than federal-provincial problems. How do we harmonize and evaluate, *ex ante*, regionalized policies, decentralized fiscal and monetary instruments, and so on? This new priority has implications for evaluative forums and actors.

A variety of other comments are in order. I agree with Professor Maslove on the importance of examining tax expenditures. In addition, I would add that it is important to develop methods and forums for the examination of non-financial decisions which do not immediately give rise to expenditures. Examples that come to mind are GATT, the Autopact, regulation, I.T.&C. industrial policy, and firm location incentives. These developments are ongoing and are being stressed in Quebec.

Another problem is the relative slowness and smallness of the impacts of monetary and fiscal policies. In my view, these give rise to a need for structural sectoral policies and choices which involve discriminations, compensations, concensus and, hence, new forums. Also, the length and time frame for the examination of these discussions is important, given the identification, response and impact lags. It seems to me that the federal government could have a longer term horizon, and that the provinces could concentrate on the medium and shorter terms. The short term, however, brings more acute conflicts.

I agree with the need for further information, explanation, and interpretation of policies. In addition, it is important to differentiate and specialize institutions, to recognize the self-interest of politicians, bureaucrats, and special interest groups, and to identify gainers and losers. It is important to provide for log-rolling and to allow for ultimate decisions to be made in Parliament, but after going through a more thorough amd more useful evaluation process. A *sine qua non* of this is, in my view, an improvement in the inside and outside research capabilities of the House, Senate, opposition parties and other agencies. The recent experiment of the Bureau de Recherche sur la Rémuneration, in submitting to an outside panel of academics its data base that it proposes to use for negotiations with civil servants in Quebec, is a positive and interesting step in this direction.

I would also argue that we should place more emphasis on *ex ante* rather than *ex post* evaluations, and would suggest that some *ex ante* discussion and evaluation of a few programs should be undertaken every year.

Finally, I wish to conclude with a few comments on Professor Gillespie's paper. The paper is a very interesting one, but I would add that evaluation, in addition to the allocation, distribution, and stabilization perspectives should be more explicit on the international, urban-regional, technological, ecological, value, and growth implications of different programs. The message that we get from his argument is to decentralize to better reflect diversity and to make for better evaluations of programs. In his table of variables to include in the assessment, I would add the impact on potential autonomous growth of the province or region, and the flexibility of the proposed policy in light of changing circumstances. Finally, the calculation of costs and benefits should be complete in that it should encompass both public and private costs.

REFERENCES

Aaron, Henry and McGuire, Martin. 1970. "Public Goods and Income Distribution." *Econometrica* 38 (November): 907-20.

Abramovitz, Moses and Eliasberg, Vera. 1953. "Governmental Economic Activity: The Trend of Public Employment in Great Britain and the United States." *American Economic Review* 43(May): 203-15.

Adams, Ian; Cameron, Williams; Hill, Brian; and Penz, Peter. 1971. *The Real Poverty Report*. Edmonton: Hurtig.

André, Christine and Delorme, Robert. 1978. "The Long-Run Growth of Public Expenditure in France." *Public Finance* 33, no. 1-2: 42-67.

Atcheson, J.; Cameron, D.; and Vardy, D. 1974. "Regional and Urban Policy in Canada." Monograph II for the Working Group on Regional and Urban Policy Analysis in Canada. Mimeo. Ottawa.

Aucoin, Peter. 1979 (forthcoming). "Public Policy Theory and Analysis." In G.B. Doern and P. Aucoin (eds.), *Canadian Public Policy: Organization, Process and Management*. Toronto: Macmillan.

Auld, Douglas A.L. 1969. "Fiscal Policy Performance in Canada 1957-1967." *Public Finance* 24, no. 3: 427-36.

Barnard, A.; Butlin, N.G.; and Pincus, J.J. 1977. "Public and Private Sector Employment in Australia, 1901-1974." *Australian Economic Review* (1st quarter): 43-52.

Baumol, William J. 1969. "On the Discount Rate for Public Projects." In *The Analysis and Evaluation of Public Expenditures: The PPB System,* Compendium of papers submitted to the Joint Economic Committee, pp. 489-504. Washington, D.C.: Government Printing Office.

Behn, Robert D. 1977. "The False Dawn of Sunset Laws." *The Public Interest* 49 (Fall): 103-18.

Bell, Daniel. 1960. *The End of Ideology*. Glencoe, Ill.: Free Press.

Bird, Richard M. 1978. "The Growth of the Public Service in Canada." In David K. Foot (ed.), *Public Employment and Compensation in Canada: Myths and Realities,* pp. 19-44. Toronto: Buterworth & Co. for the Institute for Research on Public Policy.

Bird, R.M. and Slack, N.E. 1978. *Residential Property Tax Relief in Ontario*. Toronto: Ontario Economic Council.

Bishop, George A. 1961. "The Tax Burden by Income Class, 1958." *National Tax Journal* 14 (March): 41-58.

Borins, Sandford F. 1978. "Pricing and Investment in a Transportation Network: The Case of Toronto Airport." *Canadian Journal of Economics* 11 (November): 680-700.

Brewis, T.N. 1969. *Regional Economic Policies in Canada*. Toronto: Macmillan.

Bromley, Daniel W. 1976. "Economics and Public Decisions: Roles of the State and Issues in Economic Evaluation." *Journal of Economic Issues* 10 (December): 811-38.

Buchanan, James. 1969. *Cost and Choice: An Inquiry in Economic Theory.* Chicago: Markham.

Buchanan, James M. 1970. *The Public Finances,* 3d ed. Homewood, Ill.: Irwin.

Bucovetsky, Meyer W. 1979a. "Government as Indirect Employer." In Meyer W. Bucovetsky (ed.), *Studies in Public Employment and Compensation,* pp. 29-63. Toronto: Butterworth & Co. for the Institute for Research on Public Policy.

Bucovetsky, Meyer W. (ed.) 1979b. *Studies in Public Employment and Compensation.* Toronto: Butterworth & Co. for the Institute for Research on Public Policy.

Caiden, Naomi and Wildavsky, Aaron. 1974. *Planning and Budgeting in Poor Countries.* New York: Wiley.

Cameron, D.; Emerson, D.L.; and Lithwick, N.H. 1974. "The Foundations of Canadian Regionalism." Monograph I for the Working Group on Regional and Urban Policy Analysis in Canada. Mimeo. Ottawa.

Canada. 1968. Parliament. House of Commons. *Debates.* 28th Parliament, 1st Session, p. 68.

Canada. 1971. Parliament. Senate. Special Committee on Poverty. *Poverty in Canada.* Ottawa: Information Canada.

Canada. 1977. "Statement by Mr. Macdonald on the Provincial Economic Accounts" and "Methodological Details of Calculations Appearing in Mr. Macdonald's Comments on the Provincial Economic Accounts." Ottawa: Dept. of Finance (June 6).

Canada. 1978. Parliament. House of Commons. *Debates.* 30th Parliament, 3rd Session, April 20.

Canadian Council on Social Development. 1975. *Canadian Fact Book on Poverty.* Ottawa: Canadian Council on Social Development.

Chernick, S.E. 1966. *Interregional Disparities in Income.* Economic Council of Canada, Staff Study no. 14. Ottawa: Queen's Printer.

Christofides, N.L. 1977. "The Federal Government's Budget Constraint 1955-1975." *Canadian Public Policy* 3 (Summer): 291-98.

Cloutier, J.E. 1978. "The Distribution of Benefits and Costs of Social Security in Canada, 1971-1975." Discussion Paper no. 108. Ottawa: Economic Council of Canada.

Cook, Gail C.A. 1976. *Opportunities for Choice.* Montreal: C.D. Howe Research Institute.

Co-ordination Group. 1977. Federal-Provincial Relations Office, "Preliminary Observations on the Economic Accounts of Quebec." Mimeo. Ottawa.

Curtis, D.C.A. and Kitchen, H.M. 1975. "Some Quantitative Aspects of Canadian Budgetary Policy 1953-1971." *Public Finance* 30, no. 1: 108-26.

Department of Regional Economic Expansion. 1969. ''Salient Features of Federal Regional Development Policy in Canada.'' Mimeo. Ottawa.

Department of Regional Economic Expansion. 1976. ''Climate for Regional Development.'' Mimeo. Ottawa.

Deutsch, Antal. 1968. *Income Redistribution Through Canadian Federal Family Allowances and Old Age Benefits*. Queen's University Papers in Taxation and Public Finance, no. 4. Toronto: Canadian Tax Foundation.

Diamond, J. 1977. ''The New Orthodoxy in Budgetary Planning: A Critical Review of Dutch Experience.'' *Public Finance* 32, no. 1: 56-76.

Dobell, Peter. 1977. ''The Role of Estimates Committees.'' Paper prepared for the Royal Commission on Financial Management and Accountability. Ottawa.

Dodge, David A. 1975. ''Impact of Tax, Transfer, and Expenditure Policies of Government on the Distribution of Personal Income in Canada.'' *Review of Income and Wealth* 21 (March): 1-52.

Doern, G.B. 1977. ''The Relevance and Transferability of Selective British Institutions to Canada: The Accounting Officer Concept and the Expenditure White Paper Process.'' Report to the Royal Commission on Financial Management and Accountability. Ottawa.

Doern, G.B. 1978. *Public Scrutiny of Canadian Government Expenditures: The Case for an Annual White Paper on Government Expenditure*. Ottawa: Rideau Public Policy Research Group.

Doern, G.B. and Aucoin, P. 1979 (forthcoming). *Canadian Public Policy: Organization, Process and Management*. Toronto: Macmillan.

Downs, Anthony. 1957. *An Economic Theory of Democracy*. New York: Harper & Row.

Eckstein, Otto. 1961. ''A Survey of the Theory of Public Expenditure Criteria.'' In James Buchanan (ed.), *Public Finances: Needs, Sources and Utilization,* pp. 439-94. New York: National Bureau of Economic Research.

Economic Council of Canada. 1965. *Second Annual Review: Towards Sustained and Balanced Economic Growth,* chap. 5. Ottawa: Queen's Printer.

Economic Council of Canada. 1968. *Fifth Annual Review: The Challenge of Growth and Change*. Ottawa: Queen's Printer.

Economic Council of Canada. 1970. *Performance and Potential, Mid-1950's to Mid-1970's*. Ottawa: Information Canada.

Economic Council of Canada. 1971. *Eighth Annual Review: Design for Decision-Making: An Application to Human Resources Policies*. Ottawa: Information Canada.

Economic Council of Canada. 1973. *Tenth Annual Review: Shaping the Expansion*. Ottawa: Information Canada.

Economic Council of Canada. 1974. *Eleventh Annual Review: Economic Targets and Social Indicators*. Ottawa: Information Canada.

Economic Council of Canada. 1975. *Twelth Annual Review: Options for Growth*, chap. 2. Ottawa: Information Canada.

Economic Council of Canada. 1976. *Thirteenth Annual Review: The Inflation Dilemma*. Ottawa: Supply and Services Canada.

Economic Council of Canada. 1977. *Living Together: A Study of Regional Disparities*. Ottawa: Supply and Services Canada.

Feldstein, Martin S. 1964a. ''The Social Time Preference Discount Rate in Cost Benefit Analysis.'' *Economic Journal* 74 (June): 360-79.

Feldstein, Martin. S. 1964b. ''Net Social Benefit Calculation and the Public Investment Decision.'' *Oxford Economic Papers* (New Series) 16 (March): 114-31.

Fisher, Anthony C.; Krutilla, John V.; and Cicchetti, Charles J. 1972. ''Alternative Uses of Natural Environments: The Economics of Environmental Modification.'' In John V. Krutilla (ed.), *Natural Environments: Studies in Theoretical and Applied Analysis*, pp. 18-53. Washington, D.C.: Johns Hopkins University Press for Resources for the Future.

Foot, David K. (ed.) 1978. *Public Employment and Compensation in Canada: Myths and Realities*. Toronto: Butterworth & Co. for the Institute for Research on Public Policy.

Foot, David K. and Thadaney, Percy. 1978. ''The Growth of Public Employment in Canada: The Evidence from Taxation Statistics, 1946-1975.'' In David K. Foot (ed.), *Public Employment and Compensation in Canada: Myths and Realities*, pp. 45-62. Toronto: Butterworth & Co. for the Institute for Research on Public Policy.

Foot, David K.; Scicluna, Edward; and Thadaney, Percy. 1978a. ''The Seasonality of Government Employment in Canada.'' In David K. Foot (ed.), *Public Employment and Compensation in Canada: Myths and Realities*, pp. 93-105. Toronto: Butterworth & Co. for the Institute for Research on Public Policy.

Foot, David K.; Scicluna, Edward; and Thadaney, Percy. 1978b. ''The Growth and Distribution of Federal, Provincial and Local Government Employment in Canada.'' In David K. Foot (ed.), *Public Employment and Compensation in Canada: Myths and Realities*, pp. 63-92. Toronto: Butterworth & Co. for the Institute for Research on Public Policy.

Friedman, M. 1962. *Capitalism and Freedom*. Chicago: University of Chicago Press.

Fuchs, Victor R. 1968. *The Service Economy*. New York: Columbia University Press.

Fuchs, Victor R. 1977. ''The Service Industries and U.S. Economic Growth since World War II.'' Working Paper no. 211, mimeo. Palo Alto, Calif.: National Bureau of Economic Research.

Fullerton, Douglas. 1977. ''Ottawa Should Get Back in Touch with ''the Real World''.'' *Toronto Star* (October 24).

Gillespie, W. Irwin. 1965. "Effect of Public Expenditures on the Distribution of Income." In Richard A. Musgrave (ed.), *Essays in Fiscal Federalism,* pp. 122-86. Washington, D.C.: The Brookings Institution.

Gillespie, W. Irwin. 1966. *The Incidence of Taxes and Public Expenditures in the Canadian Economy.* Studies of the Royal Commission on Taxation, no. 2. Ottawa: Queen's Printer.

Gillespie, W. Irwin. 1973. "The Federal Budget as Plan, 1968-1972." *Canadian Tax Journal* 21 (January-February): 64-84.

Gillespie, W. Irwin. 1976. "On the Redistribution of Income in Canada." *Canadian Tax Journal* 24 (July-August): 417-50.

Gillespie, W. Irwin. 1977a. "The Redistribution of Income in Canada." Mimeo. Ottawa.

Gillespie, W. Irwin. 1977b. "Postwar Canadian Fiscal Policy Revisited, 1945-1975." Mimeo. Ottawa.

Gillespie, W. Irwin. 1978. *In Search of Robin Hood: The Effect of Federal Budgetary Policies During the 1970s on the Distribution of Income in Canada.* Montreal: C.D. Howe Research Institute, Canadian Economic Policy Committee.

Gillespie, W. Irwin and Kerr, Richard. 1977. *The Impact of Federal Regional Economic Expansion Policies on the Distribution of Income in Canada.* Discussion Paper no. 85. Ottawa: Economic Council of Canada.

Ginzberg, Eli; Hiestand, D.L.; and Reubens, B.G. 1965. *The Pluralistic Economy.* New York: McGraw-Hill.

Gordon, H. Scott. 1966. "A Twenty Year Perspective: Some Reflections on the Keynesian Revolution in Canada." in S.F. Kaliski (ed.), *Canadian Economic Policy Since the War,* pp. 23-46. Montreal: Canadian Trade Committee.

Gordon, David M. 1972. "Taxation of the Poor and the Normative Theory of Tax Incidence." *American Economic Review* 62 (May): 319-28.

Gramlich, Edward M. 1968. "Measures of the Aggregate Demand Impact of the Federal Budget." In Wilfred Lewis, Jr. (ed.), *Budget Concepts for Economic Analysis,* pp. 110-27. Washington, D.C.: The Brookings Institution.

Green, C. and Cousineau, J.M. 1976. *Unemployment in Canada: The Impact of Unemployment Insurance.* Economic Council of Canada Research Study. Ottawa: Supply and Services Canada.

Grubel, Herbert G.; Maki, Dennis; and Sax, Shelley. 1975. "Real and Insurance-Induced Unemployment in Canada." *Canadian Journal of Economics* 8 (May): 174-91.

Gunderson, Morley. 1979. "Professionalization of the Canadian Public Sector." In Meyer W. Bucovetsky (ed.), *Studies in Public Employment and Compensation,* pp. 81-123. Toronto: Butterworth & Co. for the Institute for Research on Public Policy.

Harberger, Arnold C. 1971. ''Three Basic Postulates for Applied Welfare Economics: An Interpretive Essay.'' *Journal of Economic Literature* 9 (September): 785-97.

Harris, C.P. 1975. ''Financing Local Government in Australia.'' Reprint no. 6. Canberra: Australian National University, Centre for Research on Federal Financial Relations.

Hartle, Douglas G. 1978. *The Expenditure Budget Process in the Government of Canada*. Canadian Tax Papers no. 60. Toronto: Canadian Tax Foundation.

Hartman, Robert A. 1978. ''Multiyear Budget Planning.'' In J.A. Pechman (ed.), *Setting National Priorities: The 1979 Budget,* pp. 307-14. Washington, D.C.: The Brookings Institution.

Haveman, Robert H. 1965. *Water Resource Investment and the Public Interest*. Nashville, Tenn: Vanderbilt University Press.

Haveman, Robert H. 1969. ''Evaluating Public Expenditures under Conditions of Unemployment.'' In *The Analysis and Evaluation of Public Expenditures: The PPB System,* Compendium of papers submitted to the Joint Economic Committee, pp. 547-64. Washington, D.C.: Government Printing Office.

Health and Welfare Canada. 1977. *The Distribution of Income in Canada: Concepts, Measures and Issues*. Social Security Research Report no. 4. Ottawa: Health and Welfare Canada, Policy Research and Long Range Planning Branch (Welfare).

Heclo, H. and Wildavsky, A. 1979 (forthcoming). *The Private Government of Public Money: Community and Policy In British Public Administration*. 2d ed. Berkeley: University of California Press.

Helliwell, John; Sparks, Gordon; and Frisch, Jack. 1973. ''The Supply Price of Capital in Macroeconomic Models.'' In Alan A. Powell and Ross A. Williams (eds.), *Econometric Studies of Macro and Monetary Relations,* pp. 261-83. Amsterdam: North Holland.

Henderson, D.W. and Rowley, J.C.R. 1977. *The Distribution and Evolution of Canadian Family Incomes, 1965-1973*. Discussion Paper no. 91. Ottawa: Economic Council of Canada.

Hettich, Walter. 1971. *Why Distribution Is Important: An Examination of Equity and Efficiency Criteria in Benefit-Cost Analysis*. Economic Council of Canada Special Study no. 19. Ottawa: Information Canada.

Hettich, Walter. 1976. ''Distribution in Benefit-Cost Analysis: A Review of Theoretical Issues.'' *Public Finance Quarterly* 4 (April): 123-50.

Hiestand, D.L. 1977. ''Recent Trends in the Not-for-Profit Sector.'' In *Research Papers Sponsored by the Commission on Private Philanthropy and Public Needs,* vol. I. Washington, D.C.: Department of the Treasury.

Hodgetts, J.E. and Dwivedi, O.P. 1974. *Provincial Governments as Employers*. Montreal: McGill-Queen's University Press.

Horner, Keith and MacLeod, Neil. 1975. *Changes in the Distribution of Income in Canada*. Staff Working Papers, no. Z-7507. Ottawa: Health and Welfare Canada, Policy Research and Long Range Planning (Welfare).

Jackson, Peter M. 1979. "Comparative Public Sector Growth: A United Kingdom Perspective." In Meyer W. Bucovetsky (ed.), *Studies in Public Employment and Compensation,* pp. 125-59. Toronto: Butterworth & Co. for the Institute for Research on Public Policy.

Jenkins, Glenn P. 1972. "The Measurement of Rates of Return and Taxation from Private Capital in Canada." In W.A. Niskanen *et al.* (eds.), *Benefit Cost and Policy Analysis,* pp. 211-45. Chicago: Aldine.

Johnson, James A. 1968. *The Incidence of Government Revenues and Expenditures*. A Study prepared for the Ontario Committee on Taxation. Toronto: Queen's Printer.

Jump, G.V. and Wilson, T.A. 1974. "Canadian Fiscal Policy: 1973-74." *Canadian Tax Journal* 22 (January-February): 47-57.

Jump, G.V. and Wilson, T.A. 1975. "Macro-Economic Effects of Federal Fiscal Policies: 1974-1975." *Canadian Tax Journal* 23 (January-February): 55-62.

Jump, G.V. and Wilson, T.A. 1976. "Fiscal Policy in Recession and Recovery, 1975-1976." *Canadian Tax Journal* 24 (March-April): 132-43.

Kesselman, Jonathan R. 1977. "Non-Business Deductions and Tax Expenditures in Canada: Aggregates and Distributions." *Canadian Tax Journal* 25 (March-April): 160-79.

Kroeker, Hal. 1978. *Accountability and Control: The Government Expenditure Process*. Montreal: C.D. Howe Research Institute.

Lithwick, N.H. 1971. *Urban Poverty*. Urban Canada Problems and Prospects: Research Monograph no. 1. Ottawa: Central Mortgage and Housing Corporation.

Love, Roger and Wolfson, Michael C. 1976. *Income Inequality: Statistical Methodology and Canadian Illustrations*. Statistics Canada, Cat. No. 13-559. Ottawa: Information Canada.

Manga, P. 1978. *The Income Distribution Effect of Medical Insurance in Ontario*. Occasional Paper No. 6. Toronto: Ontario Economic Council.

Manser, R. 1975. "Public Policies in Canada: A Development Perspective." Paper presented to the Canadian Political Science Association. Edmonton.

Marglin, Stephen A. 1963. "The Social Rate of Discount and the Optimal Rate of Investment." *Quarterly Journal of Economics* 77 (February): 95-111.

Marglin, Stephen A. 1967. "Superstructure." In *Public Investment Criteria,* pp. 15-39. Cambridge, Mass.: MIT Press.

Margolis, J. 1969. "Shadow Prices for Incorrect or Nonexistent Market Values." In *The Analysis and Evaluation of Public Expenditures: The PPB System,* Compendium of papers submitted to the Joint Economic Committee, pp. 533-46. Washington, D.C.: Government Printing Office.

Maslove, Allan M. 1973. *The Pattern of Taxation in Canada.* Economic Council of Canada Research Study. Ottawa: Information Canada.

Maslove, Allan M. 1975. "Indicators and Policy Formation." *Canadian Public Administration* 18 (Fall): 474-85.

Matador, Bruce. 1977. "A Case Study: The Proposed Insulation Requirements for Ceilings and Opaque Walls." Ottawa: Treasury Board.

McInnis, R. Marvin. 1968. "The Trend of Regional Income Differentials in Canada." *Canadian Journal of Economics* 1 (May): 440-70.

Michelson, Stephen. 1970. "The Economics of Real Income Redistribution." *Review of Radical Political Economics* (Spring): 75-86.

Mishan, E.J. 1971. *Cost Benefit Analysis.* London: Unwin University Books.

Meerman, Jacob P. 1974. "The Definition of Income in Studies of Budget Incidence and Income Distribution." *Review of Income and Wealth* 20 (December): 515-22.

Musgrave, Richard A. 1959. *The Theory of Public Finance.* New York: McGraw-Hill.

Musgrave, Richard A. 1969. "Cost-Benefit Analysis and the Theory of Public Finance." *Journal of Economic Literature* 7 (September): 797-806.

National Council of Welfare. 1976. *The Hidden Welfare System.* Ottawa: National Council of Welfare.

National Energy Board. 1977. *Reasons for Decisions, Northern Pipelines,* vol 2. Ottawa: Supply and Services Canada.

Nienaber, Jeanne and Wildavsky, Aaron. 1973. *The Budgeting and Evaluation of Federal Recreation Programs; Or Money Doesn't Grow on Trees.* New York: Basic Books.

Niskanen, William A., Jr. 1971 *Bureaucracy and Representative Government.* Chicago: Aldine, Atherton.

Okun, Arthur M. and Teeters, Nancy H. 1970. "The Full Employment Surplus Revisited." *Brookings Papers on Economic Activity* 1, no. 1: 77-110.

Ontario. 1977. "Federal Fiscal Redistribution within Canada." In *Ontario Budget, 1977.* Toronto: Queen's Printer.

Ontario Economic Council. 1976. *Issues and Alternatives 1976. Social Security.* Toronto: Ontario Economic Council.

Paquet, Gilles. 1971. "Social Science Research as an Evaluative Instrument for Social Policy." *Social Science and Social Policy*, G.E. Nettler and K. Krotki (eds.), pp. 51-66. Edmonton: The Human Resources Research Council of Alberta.

Peacock, Alan. 1974. "The Treatment of Government Expenditure in Studies of Income Re-distribution." In W.L. Smith and J.M. Culbertson (eds.), Public Finance and Stabilization Policy: Essays in Honor of Richard Musgave, pp. 151-67. Amsterdam: North Holland.

Peacock, Alan. 1977. "Giving Economic Advice in Difficult Times." *The Three Banks Review* 113 (March): 3-23.

Peacock, Alan and Shannon, Robin. 1968. "The Welfare State and the Redistribution of Income." *Westminster Bank Review* (August): 30-46.

Pechman, Joseph A. 1977. *Setting National Priorities: the 1978 Budget.* Washington, D.C.: The Brookings Institution.

Peppard, Donald M., Jr. 1976. Toward a Radical Theory of Fiscal Incidence." *Review of Radical Political Economics* 8 (Winter): 1-16.

Perry, David B. 1976. "Fiscal Figures: Corporation Tax Expenditures." *Canadian Tax Journal* 24 (September-October): 528-33.

Phidd, R.W. and Doern, B.G. 1978. *The Politics and Management of Canadian Economic Policy.* Toronto: Macmillan.

Podoluk, Jenny R. 1968. *Incomes of Canadians.* 1961 Census Monograph. Cat. No. 99-544. Ottawa: Dominion Bureau of Statistics.

Prest, A.R. and Turvey, R. 1965. "Cost-Benefit Analysis: A Survey." *Economic Journal* 75 (December): 683-733.

Prince, Michael. 1979 (forthcoming). "Policy Advisory Units in Government Departments." In G.B. Doern and P. Aucoin (eds.), *Canadian Public Policy: Organization, Process and Management.* Toronto: Macmillan.

Pyhrr, Peter A. 1977. "The Zero-Base Approach to Government Budgeting." *Public Administration Review* 37 (January-February): 1-8.

Québec. 1977a. Ministère de l'Industrie et du Commerce. "Les Comptes économiques du Québec." Présentation de M. Roderique Tremblay, Ministre de l'Industrie et du Commerce. Québec: (le 25 mars).

Québec. 1977b. *Comptes économiques du Québec, Revenues et depenses, Estimations Annuelles 1961-1975.* Quebec: Editeur officiel du Québec.

Reynolds, Morgan and Smolensky, Eugene. 1977. *Public Expenditures, Taxes and the Distribution of Income: The United States, 1950, 1961, 1970.* New York: Academic Press.

Royal Commission on Government Organization. 1962. *Management of the Public Service.* Ottawa: Queen's Printer.

Royal Commission on Taxation. 1966. *Report.* Ottawa: Queen's Printer.

Sawers, Larry and Wachtel, Howard M. 1975. "Theory of the State, Government Tax and Purchasing Policy, and Income Distribution." *Review of Income and Wealth* 21 (March): 111-24.

Schick, Allen. 1966. "The Road to PPB: The Stages of Budget Reform." *Public Administration Review* 26 (December): 243-58.

Sewell, D.O. 1971. *Training the Poor.* Kingston: Queen's University, Industrial Relations Centre.

Sewell, W.R.D.; Davis, J.D.; Scott, A.D.; and Ross, D.W. 1962. *Guide to Benefit-Cost Analysis.* Resources for Tomorrow Conference, October 1961. Ottawa: Queen's Printer.

Shoup, Carl S. 1975. "Surrey's Pathways to Tax Reform—A Review Article." *Journal of Finance* 30 (December): 1329-41.

Smyth, David J. 1970. "The Full Employment Budget Surplus: Is It a Useful Policy Guide?" *Public Policy* 18 (Winter): 289-300.

Statistics Canada. 1976. *Income Distributions by Size in Canada, 1974*. Cat. No. 13-207. Ottawa: Supply and Services Canada.

Statistics Canada. 1977. *Provincial Economic Accounts—Experimental Data 1961-1976*. Cat. No. 13-213. Ottawa: Supply and Services Canada.

Surrey, Stanely S. 1973. *Pathways to Tax Reform: The Concept of Tax Expenditures*. Cambridge, Mass.: Harvard University Press.

Thomas, Paul. 1977. "Parliament and the Purse Strings." Paper prepared for the Royal Commission on Financial Management and Accountability. Ottawa.

Transport Canada. 1978. *STOL and Short Haul Air Transportation in Canada*. Ottawa: Supply and Services Canada.

Treasury Board. 1976. *Benefit-Cost Analysis Guide*. Ottawa: Information Canada.

Usher, Dan. 1975. "Some Questions About the Regional Development Incentives Act." *Canadian Public Policy* 1 (Autumn): 557-75.

Weisbrod, Burton A. 1968. "Income Redistribution Effects and Benefit-Cost Analysis." In Samuel B. Chase, Jr. (ed.), *Problems in Public Expenditure Analysis*, pp. 177-213. Washington, D.C.: The Brookings Institution.

Weisbrod, Burton. 1969. "Collective Action and the Distribution of Income: A Conceptual Approach." in *The Analysis and Evaluation of Public Expenditures: The PPB System*, Compendium of papers submitted to the Joint Economic Committee, pp. 177-200. Washington, D.C.: Government Printing Office.

Wildavsky, Aaron. 1977. "Policy Analysis Is What Information Systems Are Not." *New York Affairs* 4, no. 2. (Spring).

Wildavsky, Aaron. 1979 (forthcoming). *Speaking Truth to Power: The Art and Craft of Policy Analysis*. Boston: Little, Brown.

Will, Robert M. 1967. *Canadian Fiscal Policy, 1945-63*. Studies of the Royal Commission on Taxation, no. 17. Ottawa: Queen's Printer.

CONTRIBUTORS

Hugh Armstrong is on the faculty of Marianopolis College in Montreal.

Richard M. Bird is with the Department of Political Economy and the Institute for Policy Analysis of the University of Toronto.

Dean Crowther is Deputy Director, Program Analysis Division of the United States General Accounting Office.

James Cutt is with the School of Public Administration, University of Victoria.

G. Bruce Doern is the Director of the School of Public Administration of Carleton University.

Michael English is the Labour Member of Parliament for Nottingham West and Chairman of the General Sub-Committee of the Expenditure Committee of the British House of Commons.

David K. Foot is with the Department of Political Economy and the Institute for Policy Analysis of the University of Toronto.

W. Irwin Gillespie is on the faculty of the Department of Economics of Carleton University.

Allan M. Maslove is on the faculty of the School of Public Administration, Carleton University.

Pierre-Paul Proulx is with the Département de sciences économiques, Université de Montréal.

Harold A. Renouf is the Chairman of the Anti-Inflation Board.

Harry S. Rogers is the Comptroller General of Canada.

Aaron Wildavsky is a former Dean of the School of Public Affairs at the University of California, Berkeley.

APPENDIX

Program of the National Conference on Methods and Forums for the Public Evaluation of Government Spending, October 19-21, 1978.

THURSDAY, OCTOBER 19, 1978

6:00-7:15 pm.
Registration-Skyline Hotel

7:30 p.m.
Opening Remarks

Dr. G. Bruce Doern, Director
School of Public Administration
Carleton University

Dr. Michael Kirby,
President
Institute for Research
on Public Policy

8:00 p.m.
Keynote Address: The Public Monitoring of Public Expenditure

Harold Renouf, Chairman
Anti-Inflation Board

FRIDAY, OCTOBER 20, 1978

9:00 a.m.
The Reform of Public Expenditure Evaluation

Dr. Aaron Wildavsky,
Former Dean of the School of Public Affairs, Berkeley

10:15 a.m.
Coffee

10:45 a.m.
Parliament and Expenditure Scrutiny and Evaluation

Mr. Michael English, Chairman of the General Subcommittee of the Expenditure Committee of British House of Commons

11:15 a.m.
The Press and Public Expenditure Evaluation

Dian Cohen,
Economic Consultant and Freelance Journalist

12:00 p.m.
Luncheon

2:00 p.m.
Panel Discussion on Expenditure Scrutiny and Reform

- Hal V. Krocker, Visiting Professor
  School of Public Administration
  Dalhousie University
- W. Twaits,
  Business Council on National Issues
- Maurice LeClair,
  Secretary of the Treasury Board
- Jim Gillies,
  Member of Parliament

4:00 p.m.
Coffee

4:30 p.m.
The American Experience with Forecasting and Evaluating: What Should the Public Expect

Dean Crowther, Deputy Director Program Analysis Division, Comptroller General's Office, Washington, D.C.

5:15 p.m.
Panel Discussion

- Dr. David Slater, Director
  Economic Council of Canada
- Carl Beigie, Executive Director
  C.D. Howe Research Institute

6:00 p.m.
Reception

SATURDAY, OCTOBER 21, 1978

9:00 a.m.
Methods of Evaluating Public Expenditure: Problems and Prospects

Dr. Irwin Gillespie,
Department of Economics
Carleton University

9:45 a.m.
**Tax Expenditures:
The Other Side of Public Spending**

Dr. Allan Maslove,
School of Public Administration
Carleton University

10:30 a.m.
Coffee

11:00 a.m.
Panel Discussion

Dr. P. Proulx,
Université de Montréal
Hugh Armstrong,
Marianopolis College

12:30 p.m.
Luncheon

2:00 p.m.
Program Evaluation: What Is It? Who Should Do It?

Harry Rogers,
Comptroller General of Canada

Commentator:
Dr. James Cutt,
School of Public Administration,
University of Victoria

3:15 p.m.
Coffee

3:45 p.m.
Bureaucratic Growth

Dr. Richard Bird,
Department of Political Economy
University of Toronto

Commentator:
Brian Purchase,
Department of Treasury and Economics,
Government of Ontario

5:00 p.m.
Closing Address and Summary

Dr. G. Bruce Doern, Director
School of Public Administration
Carleton University

6:00 p.m.
Adjournment.

ORGANIZING COMMITTEE

For the School of Public Administration, Carleton University:
Dr. G. Bruce Doern, Director
Ms. Trish Donnelly, School Administrator

For the Institute for Research on Public Policy:
Dr. Michael Kirby, President
Dr. Louis Vagianos, Assistant to the President for Special Projects
Mr. Donald Wilson, Director, Seminars on Government Operations, and Director, Conference Program.

Institute for Research on Public Policy

PUBLICATIONS AVAILABLE*

February, 1980

BOOKS

Leroy O. Stone & Claude Marceau	*Canadian Population Trends and Public Policy Through the 1980's*. 1977 $4.00
Raymond Breton	*The Canadian Condition: A Guide to Research in Public Policy*. 1977 (No Charge)
Raymond Breton	*Une orientation de la recherche politique dans le contexte canadien*. 1978 (No Charge)
J.W. Rowley & W.T. Stanbury, eds.	*Competition Policy in Canada: Stage II, Bill C-13*. 1978 $12.95
C.F. Smart & W.T. Stanbury, eds.	*Studies on Crisis Management*. 1978 $9.95
W.T. Stanbury, ed.	*Studies on Regulation in Canada*. 1978 $9.95
Michael Hudson	*Canada in the New Monetary Order—Borrow? Devalue? Restructure!* 1978 $6.95
W.A.W. Neilson & J.C. MacPherson, eds.	*The Legislative Process in Canada: The Need for Reform*. 1978 $12.95
David K. Foot, ed.	*Public Employment and Compensation in Canada: Myths and Realities*. 1978 $10.95
W.E. Cundiff & Mado Reid, eds.	*Issues in Canada/U.S. Transborder Computer Data Flows*. 1979 $6.50
G.B. Reschenthaler & B. Roberts, eds.	*Perspectives on Canadian Airline Regulation*. 1979 $13.50
P.K. Gorecki & W.T. Stanbury, eds.	*Perspectives on the Royal Commission on Corporate Concentration*. 1979 $15.95
David K. Foot	*Public Employment in Canada: A Statistical Series*. 1979 $15.00

---

* Order Address: Institute for Research on Public Policy
P.O. Box 9300, Station "A"
TORONTO, Ontario
M5W 2C7

Meyer W. Bucovetsky, ed. — *Studies on Public Employment and Compensation*. 1979 $14.95

Richard French & André Béliveau — *The RCMP and the Management of National Security*. 1979 $6.95

Richard French & André Béliveau — *La GRC et la Gestion de la Sécurité nationale*. 1979 $7.95

Leroy O. Stone & Michael J. MacLean — *Future Income Prospects for Canada's Senior Citizens*. 1979 $7.95

Douglas G. Hartle — *Public Policy Decision Making and Regulation*. 1979 $12.95

Richard Bird (in collaboration with Bucovetsky & Foot) — *The Growth of Public Employment in Canada*. 1979 $12.95

G. Bruce Doern & Allan M. Maslove, eds. — *The Public Evaluation of Government Spending*. 1979 $10.95

Richard Price, ed. — *The Spirit of the Alberta Indian Treaties*. 1979 $8.95

Peter N. Nemetz, ed. — *Energy Policy: The Global Challenge*. 1979 $16.95

Richard J. Schultz — *Federalism and the Regulatory Process*. 1979 $1.50

Lionel D. Feldman & Katherine A. Graham — *Bargaining for Cities, Municipalities and Intergovernmental Relations: An Assessment*. 1979 $10.95

Elliot J. Feldman & Neil Nevitte, eds. — *The Future of North America: Canada, the United States, and Quebec Nationalism*. 1979 $7.95

Maximo Halty-Carrere — *Technological Development Strategies for Developing Countries*. 1979 $12.95

G.B. Reschenthaler — *Occupational Health and Safety in Canada: The Economics and Three Case Studies*. 1979 $5.00

David Protheroe — *Imports and Politics: Trade Decision-Making in Canada, 1968-1979*. 1980 $8.95